The Great Hudson River Brick Industry

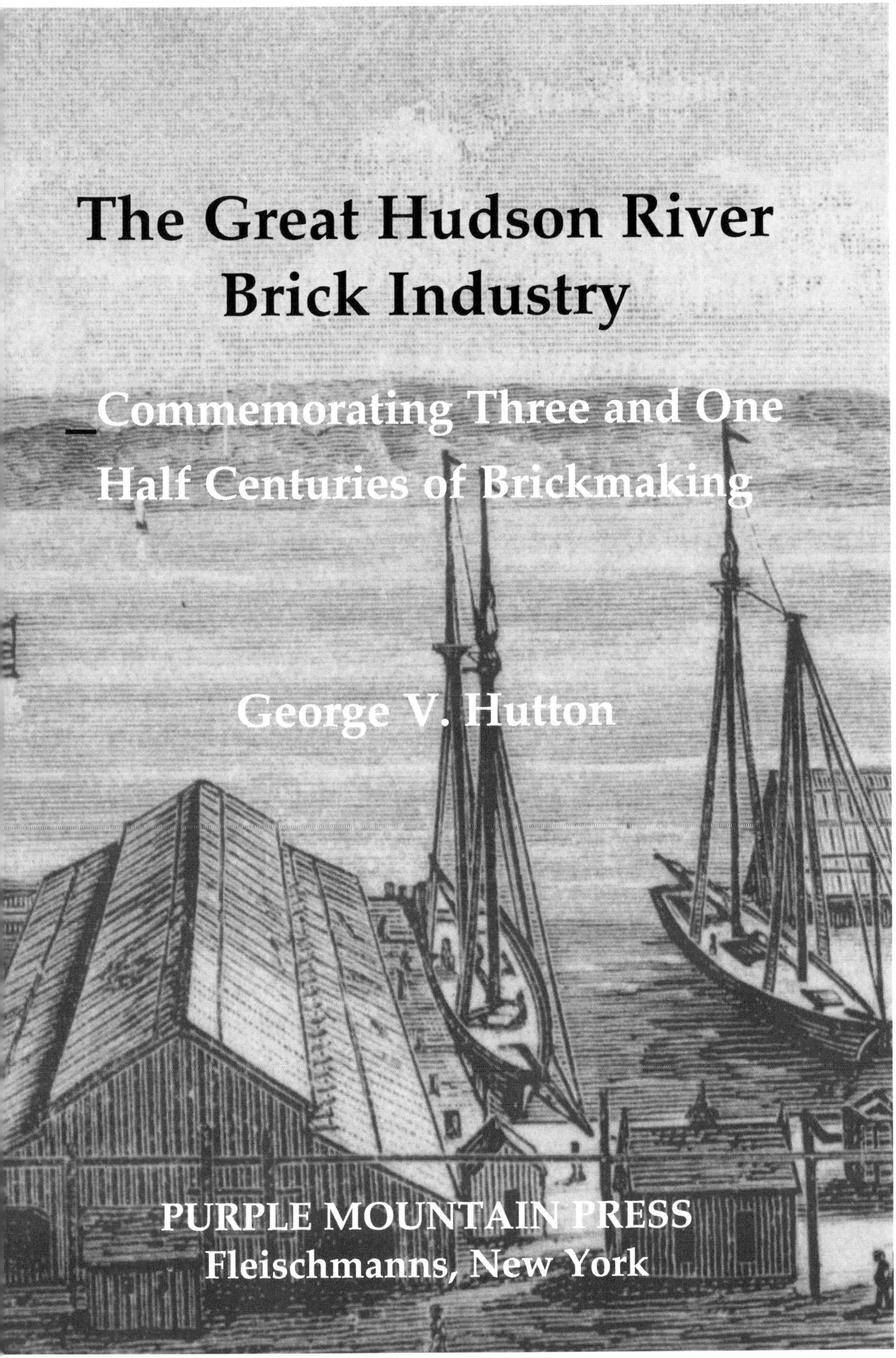

The Great Hudson River Brick Industry

Commemorating Three and One Half Centuries of Brickmaking

George V. Hutton

PURPLE MOUNTAIN PRESS
Fleischmanns, New York

The Great Hudson River Brick Industry:
Commemorating Three and One Half Centuries of Brickmaking

First Edition 2003

ColorPage Publishing
71 Tenbroeck Ave.
Kingston, NY 12401
845-331-7581
sales@colorpageonline.com
www.colorpageonline.com

ISBN 1-930098-52-9

Library of Congress Control Number 2003112678

Maps by George V. Hutton; computer assistance by Patrick Hutton
Frontispiece and title page print courtesy of Herbert L. Schultz
Cover design by Barbara Hastie

Manufactured in the United States of America on acid-free paper.

5 4 3

Table of Contents

Dedicated to the memory of
William Hutton
1926-2000
with whom this writer spent many memorable
summers at the Hutton Company brick plant.

INTRODUCTION

MANY HAVE FORGOTTEN the very great need that once existed for Hudson River bricks. The singular success of brickmakers in fulfilling that need created an industry that once dominated the Hudson Valley. This is the story of the powerful forces of urbanization in New York City that required the products of the river's brick industry and the means of producing those products for a period of three and one half centuries.

In 1910, at the end of the greatest period of growth in the history of New York City, the principal journal of the American structural clay products industry, *Brick*, stated that Chicago and the Hudson River were the largest brick manufacturing centers in the world, the latter being the largest of all. There was nothing parochial about that judgment, insofar as that journal was published in Chicago. It was for good reason that the world's largest brick industry was located on the Hudson River, considering that New York City experienced the greatest growth of any city in the world between 1880 and 1920. The prodigious growth of New York City was very much dependent on Hudson River bricks. These were the zenith years of the brickmaking on the Hudson River.

This account deals with the inception of Hudson River brickmaking in the seventeenth century and follows its development through the greatest period of its growth, the post-Civil War nineteenth-century years and the first decade of the twentieth century. The early years of this greatest growth period were also the beginning of the life of this writing's protagonist—the manufacturer having the longest run of all of them—a duration of 115 years. That brickmaker was the Hutton Company. This company was an exemplar of nineteenth-century brickmaking practice, as well as being a leader in the twentieth-century modernization of the industry. Close attention to at least one manufacturer is also done to prevent this history from becoming too generic, with unique documentation and informed firsthand opinion made available for this writing. A considerable amount of detailed description is also provided for a number of the other manufacturers. References to markets and manufacturing practices outside of New York City and the Hudson River region are provided when necessary, allowing for a better understanding of the local industry.

This publication coincides with the date of termination of brick manufacturing on the Hudson River, thereby commemorating three and one half centuries of brickmaking.

This writing notes the criticism by least one well-informed contemporary expert with respect to certain practices of the Hudson River brick industry at the turn of the nineteenth century. That criticism is cited herein, along with analysis of its applicability. This process led to a comparative examination of certain nineteenth- and early twentieth-century manufacturing practices, specifically relating to productivity and cost—a subject that is problematical to research due to the scarcity of specific information. There was some luck in picking up anecdotal data from various sources, thereby allowing reasonable analysis of the applicability of that criticism. At any rate, such criticism does add great inducement to the chase for information. It also contributes to the process of putting oneself in a situation of being a manufacturer who is faced with making decisions related to a very active and troublesome market which greatly increased the risks of making expensive capital investments.

This writing includes what may be termed `business history,' insofar as the vicissitudes of the national economy and the particular market upon which this industry was dependent do provide moments of real drama that give particular interest to the history of Hudson River brickmaking.

This history also represents a final opportunity to elicit first-hand contributions from individuals who were part of the industry's ownership, labor force, and union leadership. Regrettably, the numbers of such participants are rapidly dwindling. I also bring my own experiences with the Hutton Company, for which I am very grateful, to the writing. That personal contact with the life of the brickmaker has been a most valuable asset in the comprehension of this subject.

As well as showing the distribution of brickmaking along the Hudson River, the intention of the maps is to show the sheer intensity of manufacturing at locations of greatest activity, such as Haverstraw and Ulster County.

Finally, it must be stated that the obvious lack of currently available information on this subject prompted this writing to satisfy considerable current curiosity about this industry. Previously published writings, mostly out of print for decades, present real limitations regarding the completeness of coverage, in terms of time and geographical extent. Such writings were limited to a particular moment in time or to a limited production area on the Hudson River. This writing makes use of scarce firsthand resources for the description of the last five decades of the industry—an opportunity that had to be seized. This is not intended to be an exhaustive treatise on the industry—which would not be of compelling interest to the general reader, for whom this is most definitely written. The readership for this writing includes those with a casual interest in the subject as well as those with more technical readership requirements, and the effort is made to accommodate both levels of interest. This is intended to be a sufficiently detailed introduction to the subject in order to present the scope of the industry's accomplishments, as well as its considerable human interest.

It is with genuine regret that this writing is upon its completion, for it has been fascinating and absorbing beyond any expectation at its beginning—and I sincerely hope that some adequate measure of that great interest has been transmitted herein.

Some number of writers seem compelled to gracefully acknowledge their wife's forbearance in the presence of a process that often seems excessively absorbing and time consuming to someone with the objective detachment of the disengaged. I believe that I am, in all likelihood, in this category of writer.

In order to avoid excessive detail in the main body of the text, a significant amount of supplementary and technical information is included in the Appendix.

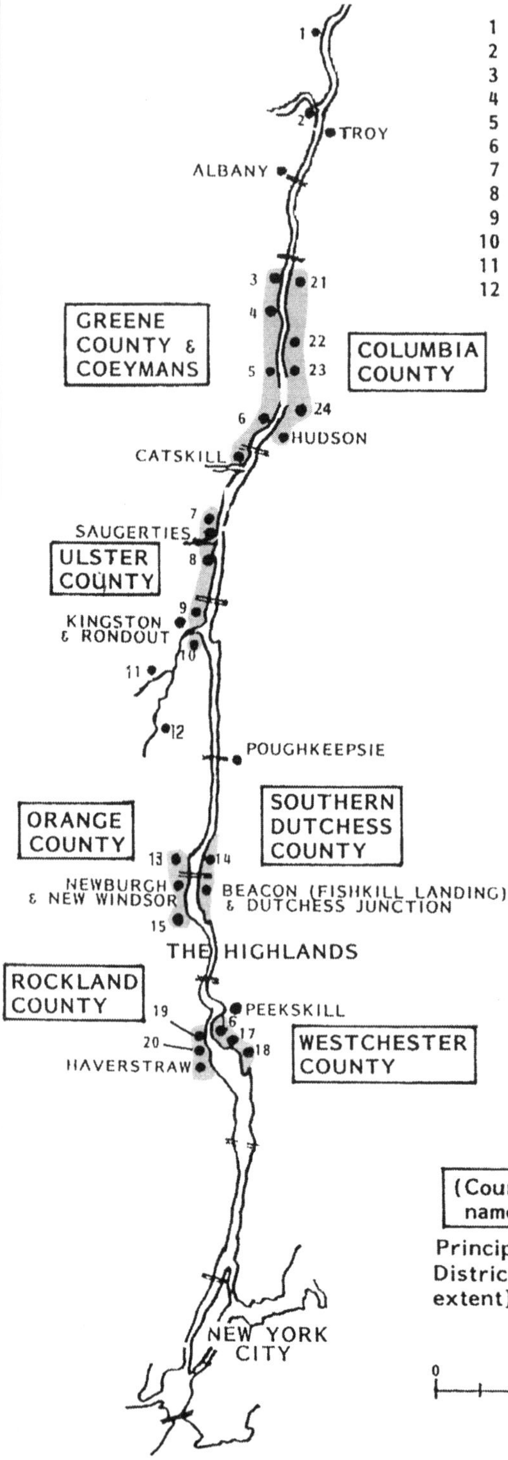

1 MECHANICVILLE
2 COHOES
3 COEYMANS
4 NEW BALTIMORE
5 COXSACKIE
6 ATHENS
7 MALDEN
8 GLASCO
9 EAST KINGSTON
10 PORT EWEN
11 ROSENDALE
12 NEW PALTZ
13 ROSETON
14 BROCKWAY & CHELSEA
15 CORNWALL
16 VERPLANCK
17 CRUGERS & MONTROSE
18 CROTON
19 STONY POINT
20 GRASSY POINT
21 SCHODACK
22 STUYVESANT
23 NEWTON HOOK
24 STOCKPORT

NUMBER OF MANUFACTURERS IN THE PRINCIPAL MANUFACTURING DISTRICTS

	1860[1]	1890[2]	1913[3]	1925[4]
Greene County &Coeymans	8[5]	8	10	7
Columbia County	0	5	6	3
Ulster County	12	19	20	19
Orange County	6	10	6	2
Southern Dutchess County	4[6]	16	15	12
Rockland County	36	44	24	10
Westchester County	29	18	6	0

Albany: A separate manufacturing and marketing entity, with eight yards supplying a local market at the end of the nineteenth century and the same number in 1925.

1 Manufactures of the U.S. - 8th Census 1860 - with exceptions as noted
2 List of Manufacturers Tributary to the New York Market 1890
3 NYS Museum Bulletin #178
4 NYS Museum Bulletin #277
5 U.S. Products of Industry - Census of 1860
6 History of Dutchess County. Hasbrouck, Editor

(County name) Principal Manufacturing Districts (Shading indicates extent)

0 10 20 30 miles

Figure 1. The Hudson River Brick Industry

CHAPTER 1
Origins of an Enterprise—Before 1865

BEFORE THE ADVENT of the enormous market for Hudson River brick of the later-nineteenth century, there was a series of necessary precedents and innovations that established the capability of the Hudson River brick industry to meet that market's demands. These included extraordinary acts of geological good fortune in locating seemingly inexhaustible supplies of clay at the edge of a very navigable waterway which led directly, in a very reasonable distance, to a voracious market for a substantial, fireproof building material that could both be produced and installed at a very reasonable cost. The raw numbers of the unprecedented growth of that market and its chaotic effects on this industry will be duly considered. The growth of New York City was highly dependent upon bricks, as were two other great metropolises going through their own surges of growth, namely Greater London and Chicago.

Brickmaking Essentials

Certain essential elements and variables that affected the brickmaking enterprise bear a very brief listing at this time. The essential raw material, clay, was consistent in quality and critical in its proximity to the market. Important implications for both of these factors appear at the end of this story. Transportation of the finished brick product to the metropolitan market was subject to price and convenience, and these elements have their own climactic impact at the end of this account. The metropolitan New York market had its own variables in the city's growth rate and in the amount of brick usage as a primary building material. This latter variable was the first to assert itself, unfavorably, to the brick business. The most critical element was the manufacturers' ability to satisfy the demand for this material. This is central to the story of the Hudson River brick industry. This story includes brick manufacturing technology, a continuing theme herein. It also includes the effects of the number of manufacturers engaged in the business and their production capabilities; these being the cause of the expansion of production, subsequent overproduction, harsh reductions in numbers of manufacturers, and times of both great confidence and tears.

Accidental encounters with this material, as in sitting next to an exposed red brick wall in a Manhattan eatery or café, gives immediacy to the enormous use of this material—it being a very safe bet that those are Hudson River brick. In fact, the odds on a wager are three to one in your favor if you wagered that the bricks are from the Hudson River. Down through the years, I have received numerous accounts from people reporting the latest sightings of Hutton Company brick alone, in the metropolitan area and one in a village near Toledo, Ohio. The brands are plain enough—each brick has it formed into the surface.

The Clay Material

There is no question about the order of events. First, nature made the clay deposits and regional and unique they were. These deposits shall be referred to herein as clay 'banks,' the term by which they were always known. These deposits are among the most extensive in America, beginning at Haverstraw (25 miles north of the George Washington Bridge) and extending 150 miles north, to Mechanicville (18 miles north of Albany), "remarkably uniform in physical properties and plastic to the point of stickiness" (Brownell). Nature was also most obliging in providing these deposits, not only in great quantity, but also in ease of access. Unlike a great many brickmaking locations in the world that require excavating pits below the surface of the ground, the Hudson River's clay banks rise in flattop terraces, or bluffs, mostly 100 to 170 feet above the river level. Within these bluffs, clay thickness varied considerably, being 18 feet at Port Ewen, 180 feet at the Rose plant (north of Newburgh) and, close by, 240 feet at the Jova plant. The average height above the river for the terraces between Glasco (2 miles south of Saugerties) and the Rondout Creek at Kingston, was 150 feet. The river face of these bluffs is steep. This feature allowed easy access to the clay by digging westward, away from the river, at ground level and higher. These clay deposits are overlaid with a considerable thickness of sand (15 to 20 feet), this sand being generally satisfactory for inclusion in the brick mixture. However, this sand is not a pure silica type, but a mixture of glacially-derived rock material.

The nature of the clay is highly 'plastic' (fine-grained, easily molded) and 'strong' (consistent quality of primarily pure clay material, with a minimum of sand and organic impurities). Such qualities were eminently suited to the molded brick that characterized the Hudson River production, but also precluded manufacturing by methods other than molding. This limitation eventually had the most serious consequences. The predominantly grey-blue color of this clay is due to its organic content. The fired color of the finished product was totally unrelated to its color in its natural condition. Yellow clays also occur where coverings of sand were nonexistent or very thin. These surface clays weathered to a yellow shade. Hudson River clays, of Pleistocene origin, have a red color when fired (red-burning), a result of iron oxide content (ferric oxide).

It must now be stated that clay is an extraordinary material made even more extraordinary by its occurrence in nature. All clays are essentially derived from the weathering of primary materials of the earth's core, with the weathering of igneous rocks furnishing the 'clay substance.' Weathering creates particles of the minutest size. An average size for clay generally is recognized as less than two microns (two one thousandths of a millimeter). Inevitably, experts disagree on this general size classification and another authority sets this size limit at four microns. The smallest sizes of clay particles approach bacterial dimensions. Hudson River clay is not of the purest type, as found in the Carolina Piedmont region, but of a material that is related in size and structure and classed as Illite-mica and Chlorite. (See Appendix H, Origins of Clay.)

The Hudson River clay deposits were transported from their original locus of creation to the bottom of a system of glacial lakes, collectively known as the Albany Lakes (extending into Lake Champlain). This glacial action also added other non-clay mineral matter to the deposits. Such is the case with the river clays, in that glacial grinding also produced a rock flour of limestone that was introduced into the clay as seasonal layers (varves), thereby adding the calcium to the clay that eventually became a heavy cross for the brick industry to bear. These clays belong to one of the two major classifications of clays according to derivation—sedimentary. When the glacier melted, a phenomenon called isostatic rebound formed the bluffs. Isostatic rebound is an upsurge caused by the removal of the glaciers' mass. Hudson River clays were derived from slow-melting glaciers, which allowed uniformly sorted beds of material to settle into the glacial lakes, as opposed to fast-melting glaciers that provide beds of coarsely sorted material. The other major classification by derivation—residual clays—became of critical importance in the mid-twentieth century.

Calcium, a well-known problem in ceramic craft and science, occurs in the Hudson River clays in quantities between 5.25 percent and 7.5 percent, the latter being by far the highest, at Coeymans (10 miles south of Albany). The calcium presented itself as a snatching back of nature's bounty that had otherwise been so very generously bestowed. At the zenith of the industry's production, in writings of 1900, a principal authority on clay and the brick industry had harsh words for Hudson River clay. Heinrich Ries stated that, "The rank of New York State among those making clay products is due rather to her location with respect to Eastern markets than to the occurrence of proper raw materials within her boundaries." This key advantage of ease of access to New York markets is a major determinant that was decisive for almost sixty years after those words were written, and insulated the Hudson Valley brick industry from the consequences of the shortcomings of the clay. Ries went on to state, "Many attempts have been made to utilize the vast deposits of clay found in the Hudson Valley for other purposes than common brick, but thus far only cases of failure are recorded." These harsh judgements of the raw materials, quite true enough, waited for over fifty years before translating into adversity for the manufacturers.

The extraordinary interaction of clay substance with water, in providing a plastic, cohesive material for molding into simple or complex shapes that could then be fired into durable products, is the basis for the great history of the usefulness of clay for several millennia. Mankind's understanding of this utility has existed for so long that it is almost genetically embedded in humans, with fired paving bricks appearing in Sumeria in 4500 BC and heroic glazed brick portrayals of lions on Babylon's Ishtar Gate in the seventh century BC. This use of brick, as an economical substitute for stone—or where the latter material was impossible to find—has persisted down through Roman and medieval times right into the third millennium AD. Brick sizes varied down through the millennia. The first approximation of the size that is familiar to us occurs in thirteenth-century Flemish bricks (8 to 9-3/4 inches long by 3-3/4 to 4-3/4 inches wide by 1-3/4 to 2-1/2 inches thick), and the English standard brick of 1685 (8-3/4 inches long by 4-1/4 inches wide by 2-1/4 inches thick) even more closely resembles modern brick sizes. In New York, an early attempt to legislate standard brick sizes was realized in 1703, with a final size established during the 1790s (7-3/4 long by 3-1/4 inches wide by 2-1/4 inches thick).[1]

Early Brickmaking

Appropriate at this time is a brief history of Hudson Valley brickmaking before 1860 and the very formative innovations preceding that date. Some of the earliest evidence of Hudson River brick usage occurred during the 1630s at Fort Orange (Albany), as shown by excavations that took place during 1970 and 1971 revealing brick foundations with dry laid bricks (without mortar). Accounts also state the use of brick in a farmhouse dating back to around 1630, in nearby Rensselaerswyck, which belonged to the greatest of all the patroons, Kiliaen Van Rensselaer, the wealthy pearl and diamond merchant from Amsterdam. Van Rensselaer also indicated local brickmaking activity in a communication to Holland in 1630, requesting skilled manpower for that purpose.

Documented history of brickmaking in Albany starts with the permission granted to Reyer Elbertsen (also spelled Elbertsz) in 1652 for brickmaking. This was in response to a concern for fire control in a growing frontier town of wood buildings that was subject to attack by Indians. A year later, Johan De Hulter arrived with a *steenbakker* (brickmaker) to establish a brickworks that his widow later sold at auction, the price payable in "merchantable beavers"—Albany was indeed on the frontier. That same kiln was resold in 1668, indicating a healthy brick trade. Manhattan builders seemed less aware of the advantages of brick, as evidenced by a 1657 ordinance:[2] "for want of stone, many wooden houses are built within this city, the one adjoining the other." Those builders were not totally oblivious, insofar as a Manhattan brickyard operated from 1647 to 1664, reputedly to provide replacements for imported bricks that were being brought from Holland. This importation itself is subject to reasonable skepticism because of the very competitively priced New Amsterdam bricks. Alan Gilbert's work on compositional analysis of early bricks also raises reasonable doubt as to the certainty of existence of consequential amounts of Dutch or Flemish imports. Presumption of Dutch origin is stronger for the yellow bricks (but lack scientific confirmation as to exact location of manufacture). Red bricks of that time have an even weaker claim to such imported origin. Cornelius Hoogeboom—already owning a brickyard in New Amsterdam since the early 1660s—operated a small brickyard in Kingston in 1665 to manufacture bricks for the purpose of building chimneys and ovens. Brick wall construction in that community waited until the nineteenth century. This kind of limited application of brick usage characterized much of the seventeenth century. In 1708, John Bryant was running a yard employing a hundred men—the largest in the entire Hudson River region—and selling the bricks by the pound! North of Manhattan, a yard appeared in Haverstraw for the first time in 1792, short lived, with another yard of brief duration appearing in 1810. During the 1860s, however, Haverstraw rapidly became the brickmaking center of the world, until the beginning of the twentieth century. In 1822, yards were still operating on Manhattan Island using early brickmaking machines that are recorded as being capable of producing twenty-five thousand brick in a twelve-hour day—a creditable production figure. Those bricks sold between five and eight dollars per thousand, a very good price that may indicate a shortage of bricks. Furthermore, those bricks were not made primarily from clay material, but from surface loam that had clay content, similar to the English 'London stock' bricks. There were two practices in making that type of English brick that was of critical importance to Hudson River bricks, to be described below.

Surprisingly, even before the 1820s, there was an active center of brickmaking at Athens, twenty-five miles south of Albany. Those bricks appeared on the New York City market at a time when sail power was the sole means of locomotion. Seven miles to the north of Athens, Coxsackie had several years of intensive brickmaking activity during the 1830s.

Technological precedents to the Hudson Valley's innovations include practices brought to this country from England during the colonial period. There were two available processes for preparing (tempering) the clay for molding. This included removal of rocks, etc. and thorough mixing by human feet or oxen hooves, with water and sand added as required. The use of oxen hooves was recorded by a U.S. Census enumerator (Products of Industries) in 1820.

A second method was the use of a 'ring pit,' a circular container in which the clay was be mixed with a wheel connected to a horse-drawn pole that was attached to an axis at the center of the pit. The very plastic, 'strong' (relatively pure) Hudson River clays required the addition of sand in order to provide adequate porosity for proper drying and firing. The tempered clay could then be forced into a mold, these originally being four-sided, open-bottom types made of hardwood, with cherry being the wood of choice.

English hand-molding practices were carried over to the colonies. Before molding began, the surface of the mold was wetted and dusted with sand ('sanded') in order to prevent adhesion of the clay. A molder shoved an eight-to-nine-pound lump of clay into the mold, thoroughly forcing it into the corners and then 'struck off' the top of the mold to remove excess material. This was strenuous manual work that had to be rapidly executed. Widely divergent production figures for a hand molder's output was fifteen hundred per day, with a maximum output of one thousand per hour by a crew of two men and two boys. An American source puts this output at five thousand brick per day for one molder with helpers. Molding was definitely a skilled workman's job, particularly with respect to obtaining a significant output. Hand molding generally followed eighteenth-century English practice, in either molding the bricks onto a 'pallet' board (this board was used to carry the bricks to the drying yard), or the clay was left in the mold and carried very carefully on edge (to prevent the brick from dropping out) to the drying yard, to be laid flat and emptied.

English practice was also followed in drying the newly molded bricks, per procedures for 'place' brick manufacture. Drying was performed on well-tamped, smooth, well-drained open-air surfaces called 'hacksteads' (suffix *-stead* from Old English meaning 'place'), from which the term 'hack,' is derived for the final step in the outdoor drying process. Newly molded bricks were laid flat, turned on edge after a short time, then built (hacked) into low, single-thickness walls (hacks), to be protected from the weather and dried for an extended period of time. The term 'hack,' as both verb and noun, was also adopted in the United States. It is from these open drying yards that the term 'brickyard,' for brick manufacturing premises, is derived. Drying removed most of the water between the clay and sand particles, rendering the 'green' brick resistant enough to damage to allow handling and stacking in the massive kilns. Hudson River kilns derived from the English 'Scotch Kiln,' a type generally described by the direction of air flow during firing as an 'updraft' kiln. With the heat source at ground level, there was a natural gravitational flow of heat, vertically, through the mass of bricks, thus raising them to the temperature required for a finished, hard product.

Mechanizing the Molding Process

The molding process was the first object of attention for mechanization and removal from the need for skilled work. A high rate of production was too dependent upon the capabilities of the individual handmolder. Early machines essentially replicated the hand-molding process, and there was gradual change in operation until the 1850s. Designs for American molding machines appeared in the late eighteenth century, with actual machine production in the United States occurring around 1820. This appears to have been a 'dry' or 'semi-dry' process using clay that had a water content ranging from five to twenty percent. The output of the initially successful machine was reported at thirty thousand bricks per day, a very good production figure.

Another supplementary molding process appeared in the early-nineteenth century, namely 'pressing' (also called repressing) bricks that had been conventionally molded by a second pressure molding that provided straight, sharp edges and smooth, dense surfaces. These were used for 'face' brick, on the exterior of wall surfaces.

In the Hudson Valley, there was keen, early interest in local development of molding machinery suitable for use with the region's clay. This was very adaptable to being forced into a mold, on account of its plasticity. Before mechanization, the hand molds had been designed to allow three bricks per mold. Early machines included a succession of designs by Nathaniel Adams, a Cornwall brickmaker (patents of 1830—Figure 2, 1836, and 1837—Figure 3). These machines, all filling four-brick molds, were powered by a manually-operated lever, a physically exhausting process. The tempered mixture was placed in a hopper that fed the press box below. Adams' machine of 1837, in modified form, was made in some quantity, finding use as far away as Chicago.

A more successful design was Hall's Improved Machine, patented in 1847 by Alfred Hall of Coxsackie, which included a vertical mixing chamber derived from English precedent, the 'pug mill.' This was essentially a vertical container that had a shaft down its center to which were attached a number of blades that both mixed the contents and forced them downward to an opening for removal into the press box. The shaft and blades were turned by horsepower. In this case, the pug mill fed directly out from the bottom into the press box for molding the brick, and this basic configuration (vertical mixing box feeding into a side-mounted press box) persisted for the next eighty years.

Integration of the pug mill into a single machine with the attached press box was a unique feature of Hall's design that was cited in his patent application. However, the press for forcing the clay into the six-brick molds was still hand operated. Even before his American patent, Hall received an English patent in 1845, an indication of the booming brick market created by the prodigious growth of Greater London.

The early machines were all predicated on the use of a very soft clay mixture that could only be obtained with the addition of an excessive amount of water. This made the newly formed brick quite soft and therefore susceptible to deformation after removal from the mold. Removal of the additional water also required extended drying times. The great breakthrough in brick machine design came in 1852 with Richard A. Ver Valen's patent for the Automatic Brick Machine that also had a vertical pug mill feeding into a side-mounted press box at the bot-

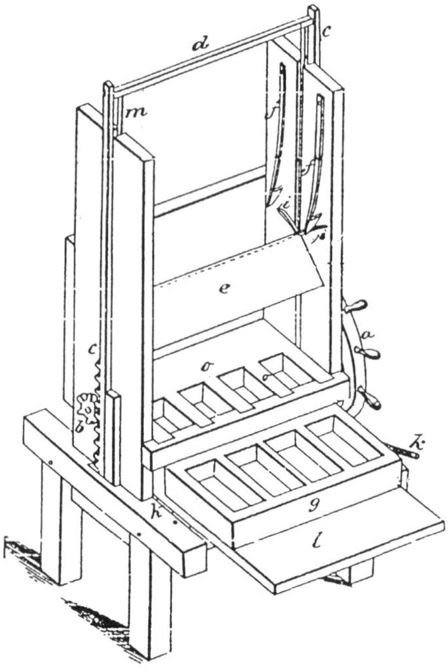

tom, but with details of mechanical advantage that allowed the use of a much stiffer clay mixture that had less water content. The design allowed for adjustments to be made, "for varying the stroke of the plunger, and thereby producing a greater or less pressure upon the clay according to its consistency or temper" (from the application filed with the U.S. Patent Office). This machine required no manual effort for any portion of its operation.

Figure 2, left. Brick Machine—Nathaniel Adams—Patented 1830. Handles for manually-operated drive wheel can be seen at lower right edge of the machine; the part labeled 'o' is a die for shaping the clay before injection into mold. (From U.S. Patent Office) Figure 3, below. Brick Machine—Nathaniel Adams—Patented 1837. (From U.S. Patent Office)

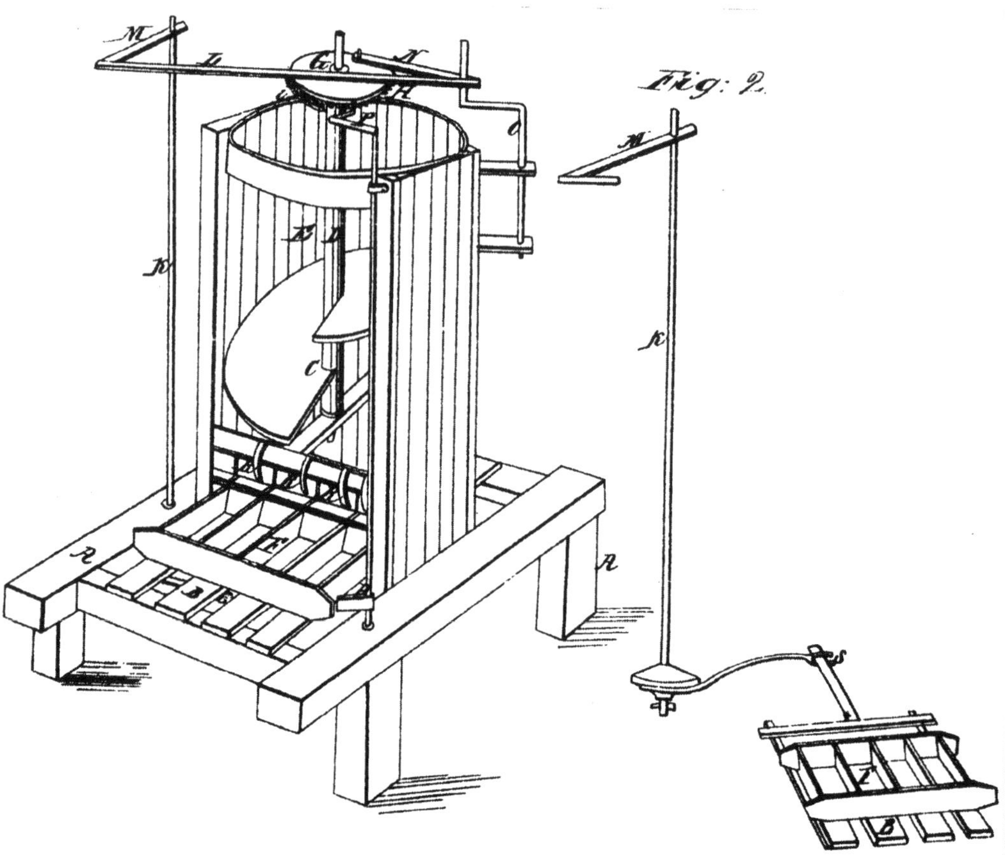

Ver Valen's family had lived in the Rockland County area since 1708, and Richard had already been involved in manufacturing stoves and plows. The illustration,[3] showing a modified successor to the original machine, depicts a large, geared horizontal wheel at the top of the pug mill for turning the vertical mixing shaft with attached knives, initially propelled by horsepower[4] and later by steam power. The illustration shows a horizontal drive shaft running across the top of the pug mill, this being turned by a steam engine at a remote location. The belt-driven wheel on the side provides the power for the press plunger and the interconnected mold feed. American machine drive belts were initially made of cowhide or horsehide. After the Civil War they were of buffalo hides because of their thickness, strength and size. These were eventually replaced by multi-ply canvas belts that were made more resistant to slipping by application of a rubber coating.

Alfred Hall, the inventor of Hall's Improved Machine, unsuccessfully sued Ver Valen for patent infringement. His suit also included users of Ver Valen's machines. Even with a number of features derived from Hall's machine (including the integral pug mill), the defendant made a sufficiently convincing argument for the uniqueness of his design. His design dominated the industry, and one hundred and fifty immediate orders plainly indicated his success.

This machine forced the clay into a mold containing six bricks with the long dimensions of the brick perpendicular to the length of the mold. The six-brick mold became a standard into the early twentieth century. The mold was placed by hand into an opening in the side of the machine

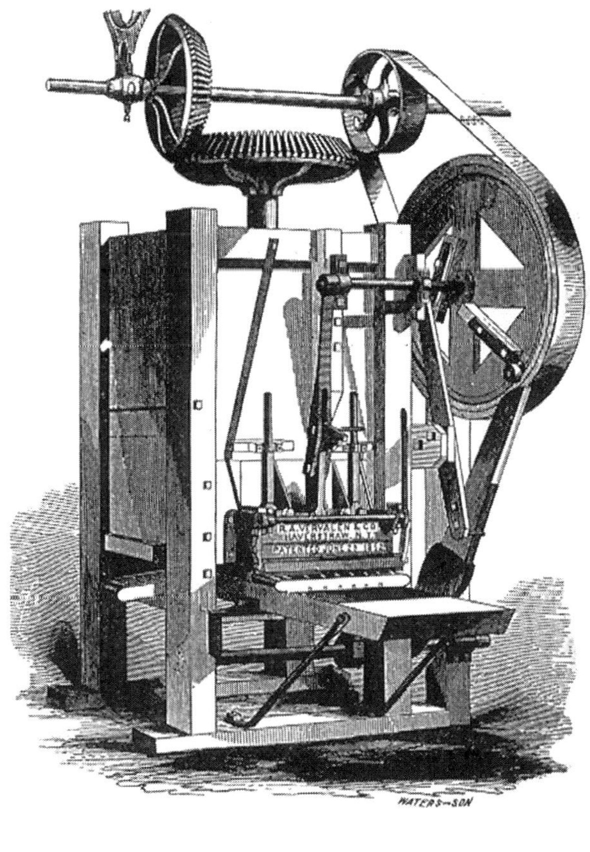

Figure 4. Brick Machine—Richard Ver Valen—Patented 1852. Horizontal continuous 'power line' drive shaft (driven by central steam engine) is at top of machine, with geared power takeoff for the machine's pug mill, and a belt-driven power takeoff for driving the molding apparatus; at the bottom of the left side of the mixing box, a mold projects from the mold insert opening. (From Cole, History of Rockland County*)*

and then mechanically introduced into the bottom of the molding box for filling. The brick was then ejected onto the molding table where the only hand operation was performed, the striking off of excess clay at the top of the mold with a long metal scraper. The timing of this invention could not possibly have been better, for it fulfilled a mounting demand for brick in the metropolitan market. All of this early inventive activity with brick machines showed the early Hudson River industry to be precocious in its development and use of new technology. McKee's authoritative writing on early American masonry stated that only a small percentage of bricks had been machine-made before 1870.

Other brick machine makers soon appeared, such as another English immigrant, John Wiles, who came to Haverstraw in 1845. After establishing himself as a Haverstraw blacksmith (that trade often being a precursor to more complex metal manufacturing such as machinery) in 1855, Wiles entered the foundry business in 1871 at Grassy Point, to become a significant maker of brick machines. His machines became widely adopted in New England, Pennsylvania, and Ohio, as well as in the Hudson River region. Incorporated in 1896 as F. J. and A. M. Wiles, the company made brick machines into the early twentieth century and also served as a place of apprenticeship, at the end of that century's first decade, for one of the most productive inventors in the history of the industry.

More Innovations

Yet another Haverstraw brick machine maker was Charles Waldron, located just north of Haverstraw (near the Peck yard). Waldron produced larger machines right down to the turn of the century.

Three additional innovations in brickmaking technology included an additive to the clay mixture that greatly reduced the time required for firing in the kiln, a critical modification to mold design, and improvements to the tempering of the clay before molding. The first two innovations vastly improved the efficiency of the practices of the time, also making it possible for the industry to meet the growing metropolitan market. A man who can be called, without exaggeration, the great original facilitator of the Hudson Valley brick industry introduced the first of these innovations. This man was James Wood, who was from a brickmaking family in England that emigrated to America in 1801. He initially began making bricks at Sing Sing on the east shore. Upon identifying rich clay-bearing lands on the west shore, Wood moved across the river to Haverstraw in 1815 to rent property from owners who were, in turn, of great significance to the industry. Those owners were the deNoyelles family who will enter into this account many times. This also established the beginning of what was to become an extraordinary practice of the Hudson Valley brick industry, right into the early years of the twentieth century.

Ownership of land for the manufacture of brick was not a requisite for being in the brick business. This practice of brickmaking on rented land, later to be extensively employed in Haverstraw, is fully explained in subsequent chapters.

There were precedents in Manhattan, where the use of common lands for the operation of brick kilns had been permitted in the eighteenth century. James Wood began production with the traditional means for tempering and hand-molding the brick. In 1829, James Wood patented an innovation that changed the world of Hudson River brickmaking. Whether it was his original invention would be a question debated for decades to come. The question as to its originality with Wood existed in the extensive prior use in England of pulverized coal as an additive to the brick material mixture (clay or loam), thereby essentially allowing the bricks to fire themselves with this internal fuel. Other similar additives such as 'soil' and 'breeze,' a mixture of cinders and coal-bearing debris from domestic ash pits, had also been used in England. Coming from a brickmaking background in that country, it could easily be surmised that Wood knew about this additive. However, the practice seems to have been totally unknown in Hudson River brickmaking practice. The revolutionary result of Wood's introduction was to reduce both the firing time and fuel consumption by fifty percent, thereby firing a kiln in six to seven days instead of thirteen to fourteen days. "This literally created the Hudson River brick industry from a slow and unlucrative business."[5] This also allowed the proper firing of bricks in the upper part of the kiln, these having previously been consistently underburned and identified by their color as 'salmon' (another term from English practice) or 'pale,' and were consigned as an inferior product to be used as filler in wall construction where no particular strength or durability was required.

Until the end of the nineteenth century, the dominant kiln fuel was wood, which was used in very large quantities. The initial coal additive to the clay mixture was anthracite, just making its appearance in the Hudson Valley after the completion of the Delaware and Hudson Canal. This was pulverized and added to the mixture in proportion of 22.5 tons of coal per 450,000 bricks, eventually to be standardized at around one percent of the mixture. The coal additive came to be known in the brick business by the name of 'culm.'

Wood proceeded to patent his invention but without Ver Valen's success in protecting it from infringement. Challenges to his patent were made on the basis of a lack of accurately stated proportions in the mix—with the impossibility of explicit statement of proportions being plainly stated in the Letters Patent, on account of the requirement of differing clay characteristics for differing amounts of coal dust. An approximate proportion was indeed stated, along with possible variations. Usage of this innovation spread so rapidly that Wood was unable to fight off infringements on his patent. Fifty years later, a chronicler wrote, "It is vain to plead that the use of coal dust in brick existed in England. It was not known here."[6]

Wood had another innovation to add, not nearly as consequential as the addition of coal dust. That innovation was a refinement of the English ring pit for tempering the clay before molding. This consisted in the insertion of short logs into a rotating horizontal shaft, which was fastened at one end to a vertical shaft at the pit's center, a horse or oxen being the motive power to pull the horizontal shaft around in circles. The shaft rotated to allow the short logs projecting from it to agitate the clay mixture. This was an approximation of the action of treading with oxen hooves and therefore a hybridization of the English ring pit.

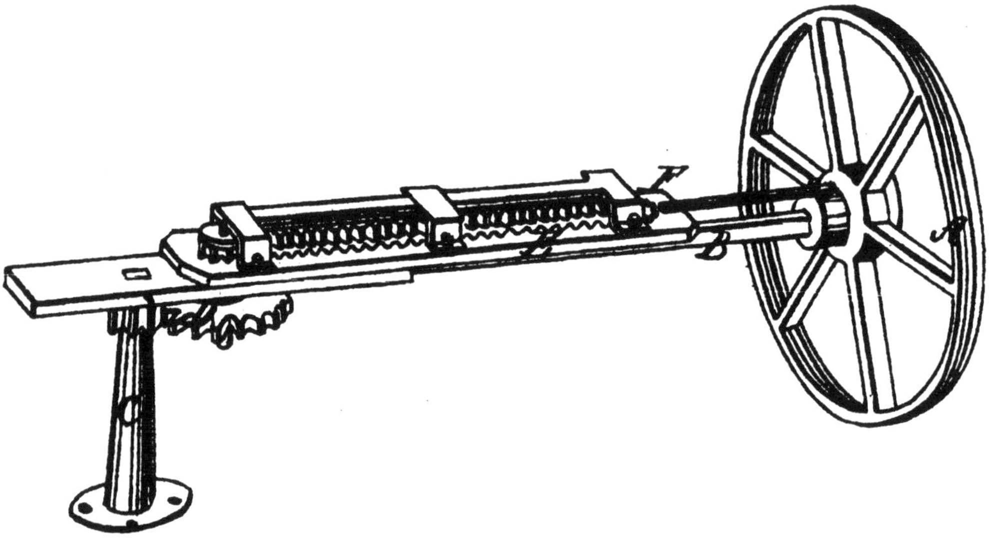

Figure 5. Tempering Wheel—Nathaniel Adams—Patented 1833. The rack mechanism is for changing radius of wheel passage through mixing pit. (From U.S. Patent Office)

More nearly similar to that English prototype was Nathaniel Adams' patented design for a metal tempering wheel. This was a spoked iron wheel with a slotted rim that provided greater interaction with the clay. A geared rack mechanism at the center of the wheel moved the wheel along the radius of the pit and continued until its direction was automatically reversed, with multiple repetitions of this movement from the outer edge of the pit toward the center and back again. Power was initially furnished by a horse in the same manner as in Woods' innovation, and then later by steam engine. This invention was widely used for the next sixty years.

The final innovation was in the development of a definitively successful mold design. The open-bottom mold presented another unresolved problem in preventing the freshly molded brick from dropping out of the mold. Before forcing in the clay, the mold had to be on a plank that allowed the freshly molded brick to be carried away from the molding table—or the filled mold had to be tipped on its side and carried away very carefully to the drying yard. The addition of a bottom to the mold, in order to form a five-sided container, had been daunted by the trapping of air in the spaces at the corners and edges of the mold bottoms, thereby preventing the proper filling of those spaces. Emptying the mold was also made difficult because of a negative pressure, or suction, after inverting the mold. A stranded crew from Tarrytown revealed the long-sought answer to this dogged problem to none other than the redoubtable James Wood. The crew proceeded to brief Wood on their yard's practice of cutting narrow slots, or 'vents,' in the mold bottom along portions of the four edges.[7] The vents allowed the escape of trapped air and broke the suction to allow emptying, but they also allowed a small amount of clay to project into the vent openings—and those small clay fins had to be removed during the drying process that followed. Molds with vented bottoms thereupon became the standard.

Another feature of mold design that had not existed until the mid-nineteenth century was the now-familiar, pronounced indentation in the flat surface of the brick, known as a 'frog.' Somehow, this English innovation from the seventeenth century had escaped transfer to the America during the colonial period. The frog had several very real advantages. It reduced the amount of material required to make a brick, as well as the drying and firing times due to the reduction in thickness and increased surface area. The frog's indentation also greatly increased the facility for handling by masons at the construction site, in reducing the brick's weight and in providing a superior means for grasping the brick by the edge of the frog's depression. Another plus for the frog—it gave improved bond in the mortar joint—and it also provided a splendid means for the manufacturer to mold the company's 'brand' (usually the maker's initials or last name) into the face of each and every brick. Thus was born one of the first uses of manufacturers' logos on mass-produced products.

The Growing Need for Bricks

As previously noted, the Hudson River brickmakers adopted the kiln type that was based upon the English Scotch Kiln, using that kiln's essential characteristics of arrangement of the bricks, firing, and air movement through the mass of bricks. Initially omitted from English practice was the construction of a permanent, thick brick masonry enclosure around the kiln space. Some of the plants later adopted this too, in a modified form. The Hudson River kiln, to be named a 'scove' kiln, dominated regional brickmaking, with its operation, advantages, and disadvantages to be discussed herein. In 1860, the scove kiln was the only available choice for achieving significant production. Other choices became available by the end of the century.

At this point, the numerous references to influences of English brickmaking practices have indicated the considerable debt owed to that country for earlier American brickmaking technology and practices. England's long history of brick manufacturing and usage began during the Roman occupation, subsequently included Saxon churches, and experienced a revitalization by Flemish refugees in the fifteenth century. The absence of economically located sources of stone necessitated the widespread use of brick in the Lowlands and North Germany. Thereafter, the manufacture and use of bricks in England rapidly gained acceptance, to the point of becoming the dominant building material around London and in many other parts of that country. Well-developed English technology and practices translated easily to America.

The need for large quantities of brick for the growth of New York City received several important early inducements. As in London, an increasing rate of population and commercial growth created a demand for a low-cost, durable, fireproof material that could also be put in place at low cost by the available construction manpower. In 1647 and 1648 Peter Stuyvesant mentioned problems in fire prevention due to wood buildings and flues made of wood and plaster. Stuyvesant came from an urbanized country already steeped in the virtues of brick masonry usage. In 1656, Stuyvesant went on to rule against flues made of wood, thereby requiring either brick or stone. The early nineteenth century shows formal regulatory activity, naming specific Lower Manhattan streets as 'fire limits' (bounded on the north by Chambers Street) within which masonry construction was required for new construction, this in 1810 and 1812, with the earliest Fire Limits Acts in 1761. The Great Fire of 1835 started on Merchant Street (nonexistent today but originally at the southern tip of Manhattan), burning for nineteen hours over fifty acres, consuming over six hundred and seventy buildings, and giving a big boost to the Hudson River brick industry during the process of rebuilding. That fire caused the bankruptcy of twenty-three of the city's twenty-six insurance companies and greatly increased efforts to build the Croton reservoir and aqueduct systems. The Great Fire indicated a large number of frame buildings—which were still in existence in the oldest part of the city—and rein-

Figure 6. Brick Mold (mid-twentieth century)—showing vent slots. (Mold in collection of Hudson River Maritime Museum; photo by author)

forced the need for fire resistant construction.

The Second Great Fire, in 1845, destroyed an additional three hundred buildings. Extension of the fire limits subsequently extended northward, as in 1847, when a new act set a fire limit from the Hudson to the East River, on a line one hundred feet north of 32nd Street. By 1866, fire resistance regulations, mandating party walls of brick or stone, were extended throughout the entire city, with the fire limit line moving north to 86th Street. Fire resistance became synonymous with brick wall construction, and the fire limits kept moving northward.

A survey of pre-Civil War brick usage, both within and related to the city, shows large amounts of brick (40 miles of brick-lined tunnel) used for the Croton Aqueduct, sewer construction (261 miles by 1872), all sidewalks of brick in the first half of the century, as well as building construction.[8] This even included large amounts of brick required for all of the principal wall construction that lay behind the cast-iron building fronts erected during the dozen years preceding the Civil War. Any collection of photographs of the city at mid-century will testify to the enormous usage of the material, not to mention the aforementioned very visible testament provided by the exposed brick walls in the cafes and restaurants that we see today.

Alternatives to brick had existed from earlier years in Manhattan. There had indeed been considerable use of stone masonry for foundations and for more principal and formal buildings. However, bricks were again performing their ancient role as an economical substitute for stone, and doing so on a very large scale. Popular tastes cause big changes in the visible world, as is the case with the large numbers of buildings that have brownstone fronts that were built in the years preceding the Civil War until around 1890. Even those buildings had great amounts of brick in the common walls running perpendicular to the street, and on rear walls, as even the most casual inspection will reveal at the present time. The list of locations for the use of great quantities of bricks includes thick brick walls bonded to the back of a stone masonry finish, and interior bearing walls and partitions. It would not be too much of a stretch to describe late-nineteenth-century New York as a 'brick city.'

The Early Manufacturers

With the existence of adequate clay resources, brickmaking technology, and demand, there remained the need for manufacturers to provide the bricks. There was no shortage of these, either. The low capital requirements for getting into the early brick business made entry relatively simple. The inestimable James Wood attested to this: "Brickmaking was a poor man's game, as it required no capital to start with."[9] This was certainly true of the hand-molding era, with the small early brick machines allowing for gradual increments of such mechanisms, initially driven by horsepower. Buildings were simple wood sheds for covering the machines and the kilns—and they essentially stayed that way until the 1920s.

Such structures were later built for the enclosure of drying operations at a limited number of yards. Horses and men initially furnished power. The investment ante was indeed low during the early days, with this to drastically increase in the twentieth century.

Two important characteristics of the Hudson River manufacturers were the widespread similarity of all the processes for making bricks and the production of a single type of brick. Choices in manufacturing processes were quite limited in the earlier years, and few alternatives subsequently became available. The physical characteristics of the clay itself imposed an important limitation on the choice of brick type to be manufactured. The similarity of manufacturing methods had essentially sound justification, in terms of the raw material and the market, all of which is discussed at some length herein.

The type of brick that was universally manufactured was the so-called 'common' brick. This was a class of brick generally designated for structural purposes, without the repressing, special finishes, grading and special handling received by 'face' bricks. Face bricks were specifically manufactured for use on exterior surfaces, although selected common bricks also fulfilled this function. Because of advantageous clay composition and an early start in the manufacture of face bricks, Philadelphia manufacturers obtained a dominant early position in the manufacture of such bricks, followed by some New Jersey makers.

There were two manufacturing regions of the brick industry along the river. By the early 1840s, most of the locations producing bricks at the end of the nineteenth century were already in production. All manufacturers below the Highlands, essentially in Rockland and Westchester Counties, became known as 'Downriver.' The 'Upriver' region provided small-scale manufacturing at various locations until its rapid growth after the Civil War. In 1860, Ulster County had small hand- and horsepowered yards employing twenty to twenty-five workers and produced two-and-a-half to three million bricks per year, per yard. Before the Civil War, Westchester and Rockland Counties were the dominant producers, and the latter brickmaking complex, by 1870, became the principal center of brickmaking into the beginning of the twentieth century. This included (from north to south) Stony Point, Grassy Point, Haverstraw, and South Haverstraw. Surprisingly early, the U.S. Census of 1820[10] showed six yards at Athens (twenty-five miles south of Albany) producing over three million bricks per year for the New York City market. Most of those yards employed three to six men and made 240,000 to 500,000 bricks per year—and one yard produced one million bricks with fourteen employees. Precocious as an early center of Hudson River brickmaking, Athens was only a minor producer in later years. The same census records show entries for Athens, under the heading of 'Manufacturing Methods': "Clay and sand ground and trod by oxen." Such were the mixing processes of that time. Coxsackie (twenty miles South of Albany in Greene County) had been both a very active and early center of Hudson River brickmaking during the 1830s. Twenty-five small yards reportedly produced fifty million bricks per season and shipped to a relatively remote New York City market by sail—a distance of more than one hundred twenty miles. Transplanted land-hungry New Englanders who entered this part of the Hudson Valley in the years after the Revolution started these yards. Coxsackie's brick industry suffered heavy damage from the national financial panic of 1837 (effects of excessive borrowing for turnpikes, canals, and the infant railroad system). This triggered a depression during the following decade ("The Hungry Forties"). One maker, Peter Hubbell, moved near Boston where he grew wealthy by manufacturing brickmaking machinery. Before 1860, one precocious Catskill brickmaker, Lee Ensign, had utilized steam power to drive his brick machines.

Central to Haverstraw brick manufacturing was the deNoyelles family, whose Huguenot ancestor acquired a farm on the bluffs overlooking the river in the mid-eighteenth century. The transformation of a rural countryside for extractive manufacturing purposes radically changed communities and the landscape, with rural activities continuing to co-exist with the brickyards for some time. During the earlier days of Haverstraw brickmaking, herds of cows were known to have wandered through drying yards that were covered with newly-molded, soft brick. A loss of production, needless to say. In locations that had brick manufacturing, the line of bluffs close to the river observably receded. As previously noted, James Wood leased a portion of the deNoyelles property for brick manufacturing in 1815. The deNoyelles later became principal manufacturers on their own lands.

The 1850 census shows a half dozen yards to the south of the Ulster County town of Saugerties and almost twenty-five at Haverstraw. The 1860 census shows that there were thirty-six yards in Rockland County, twenty-nine yards in Westchester County, twelve in Ulster County, six in Orange County, nine in Greene County, and four in southern Dutchess County—the latter being a number that very rapidly increased. The Hudson River brick industry was rapidly developing.

Other early Haverstraw manufacturers included the Peck Brick Company, which began during the 1850s and grew to be one of the most principal and progressive manufacturers on the river. The Allison yard—also to become one of the largest—had its origins from the same mid-eighteenth-century division of property from which the deNoyelles purchase was made.

The principal measure of the unit of productivity was the individual brick machine. An earlier two-machine yard at Grassy Point required a strip of land 180 feet wide and 475 feet long, with each machine producing eighteen to twenty thousand bricks per day, in accord with the conventional division of the work day for each machine crew.[11] This kind of precise calibration of the amount of land required per machine for clay extraction, molding, drying, and kiln space indicated the manufacturing demands on a production area of limited size. A rapidly growing number of manufacturers occupied the principal Rockland County production area of approximately two and one half miles in length. Such concentrated use of limited land areas would eventually exhaust economically available clay resources. Until that time, Downriver enjoyed a considerable transportation cost advantage (when limited to sail power only) over Upriver yards, where space and clay resources were far more abundant and the density of manufacturers far less. The thirty-six Rockland County manufacturers of 1860 were located along four and a half miles of the Haverstraw-Stony Point shoreline.

Westchester County had a large number of very productive yards, even before the middle of the nineteenth century, centered on the adjacent communities of Verplanck, Crugers, Montrose, and Croton—all on Haverstraw Bay. Production commenced at Croton Point in 1830, followed by the 1843 start of production at Verplanck Point (immediately south of Peekskill). The arrival of John Henry, the ropemaker from Long Island, who bought five hundred acres from Phillip Verplanck initiated this. Verplanck Point became the main center of Westchester brick production. Verplanck Point was an important property during the Revolutionary period. It was the eastern terminus of the critical crossing made by the King's Ferry and had an important position opposite the fort on Stony Point. After the Battle of Yorktown, Washington and the Comte de Rochambeau used Verplank Point as the site of a celebratory review of Continental and French Troops. The 1860 value of production of those Westchester yards (eleven of which had steam power by 1855) exceeded the value of any other brickmaking county on the river. Westchester's production was totally unique in experiencing a significant reduction by 1880—while the other counties' production rapidly expanded.

Labor Problems

Labor relations, often troubled, would be a recurring theme in the development of the industry. This included obtaining the sheer number of workers required for the rapidly increasing operations of the yards. Rockland County accounts show an apparent consensus among manufacturers regarding wages. In 1853, a simultaneous reduction in wages by a significant number of manufacturers caused all brickyard workers to strike. The result of this was a demonstration by hundreds of workers who marched from yard to yard, "breathing threats of violence."[12] Because the constraints of collective bargaining and other influences both moderating and benefiting labor relations of the late twentieth century did not yet exist, and with the community expressing feelings of insecurity, the sheriff responded by obtaining the services of a local military unit, the Rockland County Rangers. Such episodes of open conflict between management and labor were more characteristic of the Downriver industry than of the Upriver yards.

Growth in Manufacturing

The arrival of the 1860s marked the rapid growth of Upriver manufacturing. This included the start of the plant that saw the most years of operation in the history of the Hudson River brick industry. Certain manufacturers who would prominently figure in the Upriver industry were already at work. Albert Terry arrived at Steep Rocks, one mile north of Kingston Point, at age twelve. His father David had moved there in 1845 from Delaware County, immediately beginning to farm and make bricks, "with the old-fashioned one horse sweep used to grind the clay and the slow and tedious method of making brick by hand."[13] The Terry company dated its origins from 1850. Upon the death of their father in 1869, Albert and his brother Edwin went into partnership as Terry Brothers at the same location, and another plant was put in operation immediately north of East Kingston. The Steep Rocks plant operated until 1956—making it the runner-up for longevity behind the Hutton Company. Curiously coincidental is the physical juxtaposition of the two companies that had the greatest longevity of all the yards on the river.

Also in operation, by the end of the 1850s, were the Washburns. George and John T. Washburn, sons of a Haverstraw tanner, apprenticed in the brick business in Haverstraw at the ages of seventeen and fifteen, respectively. In 1859, they opened their own Haverstraw plant as Washburn Brothers, followed by a plant at Glasco (south of Saugerties) in 1867. The latter location eventually had two Washburn Brothers' yards, 'Upper' and 'Lower,' quickly became a major producer, eventually had the largest production on the river. One of those plants operated until 1949.

The ensuing four decades witnessed great changes in the Hudson River brick industry.

Notes

1 McKee. *Introduction to Early American Masonry.*
2 Stokes. *An Iconography of Manhattan Island.*
3 Figure 4. Ver Valen's Automatic Brick Machine
4 Figure 8. Horse-powered brick machine
5 Green.
6 Green.
7 deNoyelles and Figure 6.
8 O'Connor.
9 Green.
10 U.S. Census. *Products of Industry.*
11 Newell.
12 Green.
13 *Commemorative Biographical Record of Ulster County.*

Principal Sources for Chapter 1

Beecher, Raymond. *Under Three Flags.*

Brownell, Wayne. *Clays and Shales of New York State.*
(The) Carpenters Company for the City and County of Philadelphia. *Building Early America.*
Cole, Rev. David. *History of Rockland County.*
deNoyelles, Daniel. *Within These Gates.*
Dobson, Edward. *A Rudimentary Treatise on the Manufacture of Bricks and Tiles.*
Green, Frank B. *History of Rockland County.*
Hasbrouck, Frank. *History of Dutchess County.*
McKee, Harley. *Introduction to Early American Masonry.*
O'Connor, Richard. *A History of Brickmaking in the Hudson Valley.*
Stokes, Isaac. *An Iconography of Manhattan Island.*
U.S. Census. *Products of Industry*—1820, 1850, 1860.
U.S. Census. *Manufactures of the U.S.*
U.S. Patent Office. (See Appendix).

CHAPTER 2
The Great Years of Growth—1865-1897

THESE WERE YEARS of vertiginous growth in the brick business, during which the industry provided the substance required for the enormous expansion of New York City.

The first year of this period also witnessed the beginning of 115 years of brickmaking by the plant that had the longest duration in the history of the Hudson River brick business. This plant was also among the largest, at a time when there were already eighty-three manufacturers in the brick business. There were sixty-five Downriver yards (in Rockland and Westchester Counties) and eighteen Upriver, of which twelve were in Ulster County.

Upriver production was still very limited at the beginning of the 1860s. By 1870, radical change began to occur. During the 1860s, the market value of Upriver production had increased by more than five hundred percent and production in Rockland County by ninety percent, making this decade a critical time to enter the industry.[1] Even at this time, New York City had not yet experienced the prodigious surge in population that characterized the years 1880 to 1920. During that time, New York City had the largest population growth of any city in the world, outstripping the numerical population growth of the largest city in the world, Greater London, by almost forty percent (London exceeded New York City's population by one hundred fifty percent in 1880, with that lead reduced to thirty-three percent in 1920).[2]

John H. Cordts and William Hutton

Such activity in the brick business also attracted the attention of John H. Cordts and William Hutton, a resident of Rondout, who was thirty-one years old in 1865. Born in Philadelphia as the oldest of six children, William's parents were John and Margaret Dougherty Hutton. John had immigrated to the United States from Enniskillen, North Ireland. The family moved to Rondout (presumably from Philadelphia), at the mouth of the Rondout Creek near Kingston, in the mid-1850s. William's initial business venture was in groceries, in partnership with a brother. In 1857, he married Jane Kerr, daughter of Robert Kerr, a schoolteacher who immigrated to the United States in the 1830s from Castle Marvin, County Tyrone, in North Ireland. William made the first of a very large number of real estate acquisitions in 1859. These totaled over one hundred thirty purchases and an equivalent number of sales during the next thirty-seven years before his death in 1897 (averaging one transaction every other month).

The property for the brick plant was purchased in 1865, in partnership with John H. Cordts, who was ten years older than William Hutton. Cordts emigrated from Prussia (Wremen, in the province of Hanover) at age fourteen. Cordts was also initially engaged in the grocery business in New York City. He then moved to Cornwall where he learned the brick business and eventually owned a brick-yard before moving to Kingston in 1850. John Cordts also had substantial real estate operations. For twenty-five years, the enterprise was known as Cordts and Hutton. The property that they acquired was the northern portion of a royal patent granted to Moses Cantine, a Huguenot, in the first years of the eighteenth century.[3]

The entire portion of the original patent to the north of Delaware Avenue (with exception of previously sold lots) passed from Cantine ownership in 1849. Two others owned the property before the 1865 purchase by Cordts and Hutton.The north end of the property included a prominent landmark called Steep Rocks. These were bare-faced cliffs cited by that name as a well-known landmark in 1736, in minutes of the Trustees of the Freeholders and Commonality of Kingston (an entity that was as close to local self-government as the English colonial government would allow).[4] The trustees were looking for woodlots—a usage that eventually had great importance to the new brickyard of Cordts and Hutton. On the adjacent property to the north, David Terry had already been making bricks for fifteen years (with two more Kingston yards to appear north of Terry's by 1880—Nathan Nickerson and Alva Staples). That manufacturing activity, plus the rapidly growing New York City brick market, and the prospect of enormous clay resources must have made the Cantine Patent property look very attractive.

As shown by John H. Cordts's hands-on demeanor in Figure 7, he was undoubtedly a much greater presence at the brickyard than William Hutton, building his own mansion on a nearby height of ground—living over the store, so to speak. Hutton initially lived in Rondout, followed by a move up the hill to

Figure 7. John H. Cordts and Workers—1880s. Cordts is at center of the picture (with fists on hips); to Cordts's left is a young man (barefoot for working on drying yard) holding what appears to be a large metal insert for an ice cube tray—the implement for 'edging' the bricks (turning them on edge after initial drying on the flat surface); to Cordts's right is a 'molder,' bearing the long cut-off blade for scraping excess clay from the top of the mold; other men bear implements of their trade. (Courtesy of Matilda Cordts Davenport and the Cordts Family)

West Chestnut Street.

William Hutton himself, for the next twenty-five years, was virtually an absentee owner. His principal place of business was at his lumber company in Rondout—a mile from the brickyard. His building materials enterprise was one of the main businesses in that community. The 1870 U.S. Census shows him, at age thirty-six, reporting a personal estate of fifty thousand dollars (somewhat more than nine hundred thousand in present-day dollars)—not bad for a local boy with the usual limited schooling of the time.

Very plainly, Hutton intended to get rich, and his various enterprises indicate that his ambitions in that respect would not excessively rely on the brick business. His brother-in-law, Charles Van Wert, represented his interests in the brick business. Van Wert acted as superintendent and very capably held that position for a number of years. Van Wert's association with the early years of the business set a precedent for an astonishing display of the workings of fate over a century later, this to be revealed at the end of this writing.

The early brick business, not incorporated until 1897, left no records in the form of the directors' minutes that were kept after that date. Fortunately, there are two significant sources to help assemble a picture of the industry and of this brickyard at that time.

Brickmaking in the Post-Civil War Years

The U.S. Federal Census of the Products of Industry gives sufficient information to obtain critical data and to identify changes in manufacturing processes during the portion of the nineteenth century that saw rapid change in the brick business. These records provide information describing small Upriver manufacturers in 1860, each making two to three million bricks per year with thirteen to twenty-six employees. Kiln fuel was limited to wood—an average consumption of 350 cords per year—and pulverized coal (culm) was used as an additive to the brick mixture. Power was mostly furnished by men's hands and horses.[5] During the 1860s, the industry was rapidly changing from horsepower to steam power for driving the brickmaking machinery, providing a significant increase in productivity. Steam powered the essential machinery of the brickyards for the next fifty to sixty

Figure 8. Horse-Powered Brick Machine—1890s. A rare photograph of a horse-driven brick machine that is worthy of inclusion in spite of its late date and location in Plattsburgh; this is a small, portable machine used by itinerant brickmakers; note the bricks drying on the ground at lower left and the adjacent 'hack' of bricks. (Photo by Heinrich Ries; courtesy of New York State Archives).

years, until replaced by electric power.

Production had also increased at many of the yards, as well as the number of employees, including children at some of the yards. The U.S. Census of 1870 shows the Cordts and Hutton yard off to a fast start, making ten and one half million bricks with six steam-powered brick machines, thereby making this one of the largest-volume manufacturers on the river. The 1870 census shows the rest of the Ulster County yards using hand- and horsepower, converting to steam power by 1880. The 1880 Census shows that there were 115 employees at Cordts and Hutton with an average of forty-two workers per yard in the Ulster County region.

Wood persisted as the chosen kiln fuel and was used in much larger quantities—as much as eighteen hundred cords per year at Cordts and Hutton. A good portion of such wood came from the Steep Rocks portion of the property (recall that this had been cited in 1736 as a source of wood by the Trustees of the Freeholders and Commonality of Kingston). Their production of twelve million bricks per year was exceeded on the Hudson River only by the combined plants of the Washburn Brothers at Glasco (twenty million) and by two Haverstraw yards, of which the Diamond Brick Company's five combined yards had by far the largest production on the river (thirty-four million). Nathaniel Sylvester (in *History of Ulster County* of 1880) provided a production figure for Cordts and Hutton of sixteen million—plainly at odds with the U.S. Census figure. The Washburn Brothers had established their Glasco operations in 1867 while concurrently manufacturing in Haverstraw. That Downriver yard only made seven and one half million bricks in 1880. The position of the Cordts and Hutton plant (to become the Hutton Company in 1890) as a leading producer persisted for a century.

In addition, Sylvester's *History of Ulster County* made use of the Cordts and Hutton plant as the principal example of the business in the county. That history provided an adequately detailed description of "one of the largest brick-yards on the Hudson between Haverstraw and Albany." Sylvester noted, "an inexhaustible bank of clay," two lengthy kiln sheds (eight hundred feet and three hundred feet), eight brick machines with tempering wheel pits for preliminary mixing, all connected by a nine-hundred-foot-long drive shaft, this "driven by one of the best and most powerful engines used in the manufacture of brick in the state. It is of one hundred and twenty-five horsepower. . . . Three new boilers four feet in diameter and twenty-four feet in length were furnished by McEntee and Dillon, of this city, and are excellent ones." These latter names are of interest, in that they are attached to nineteenth-century artistic figures of note. Jervis McEntee, a principal member of the Hudson River school of painters, was a first cousin of Julia McEntee Dillon, a lesser-known but very capable artist noted for floral still life paintings. Julia Dillon had exhibited her works at the National Academy of Design and studied in Paris after marrying John Dillon, the owner of a foundry in which the McEntees purchased an interest, becoming McEntee and Dillon. After her husband's untimely death in 1873, Julia was listed as a partner in the foundry business. At this time she lived in New York City and made various trips to Europe (presumed not to be foundry expense account items). It was during the time of Julia's association with McEntee and Dillon that the brickyard's boilers were made. The aforementioned steam engine may have been of the Corliss type, these engines being widely used in the Hudson River brick business.

A Steam Power Primer

If, on your visit to the 1876 Philadelphia Centennial Exposition, you had seen "The Mighty Corliss" engine, you had seen the future. That fourteen-hundred-horsepower monster (sporting a thirty-foot-diameter fly wheel) was "delivered to the order of the Exhibition Commission, by which it was sought to establish the finest memento to American inventive talent."[6] The Corliss engine's innovative control of steam admission (utilizing James Watt's centrifugal governor) and its cylindrical valves enabled its precise control of steam admission to the piston cylinder—revolutionary features greatly increasing its efficiency (using forty percent less fuel than conventional engines). These were the reasons for its extraordinary success (in 1882, thirty-six patent licensees in the United States and more than two hundred forty licensees in Europe). The inventor, George H. Corliss, was one of the leading industrial innovators in the United States in the second half of the nineteenth century, holding a long list of patents on steam engine designs, and also on machinery for the cutting of teeth on massive gear wheels (his gear cutting machinery making possible the operation of the turret on the Union ironclad *Monitor* during the Civil War). Even in the first decade of the twentieth century, the Corliss engine was hailed as "without doubt the standard of economic engine performance and with its longer life is preferable for ordinary purposes."[7]

The use of this engine spread rapidly. Under patent licensing agreements, various foundries and machinery manufacturers made this engine, with significant representation in Hudson Valley brickyard usage as 'Fishkill Corliss' and 'Newburgh Corliss' engines. In Newburgh, they were manufactured by William Wright & Company. Wright was a superintendent at the Corliss Steam Engine Company works in Providence, Rhode Island. Wright maintained a querulous claim against Corliss, arguing that the principal basis for Corliss's reputation lay in a design for the famous cylindrical valve for which Wright claimed original invention (for decades treasuring a battered piece of paper upon which he claimed that he had sketched the original germ of the valve design). See Figures 10 and 11. Admittedly, a juicy bit of gossip—but Corliss also possessed a very large number of patents in addition to the valve design (including many patents for gear-cutting machinery)—and he had the great wealth and international fame that Wright so obviously envied. Nonetheless, Wright was an accomplished enginemaker, having built engines in Hartford and large pumps for the Brooklyn waterworks before moving to Newburgh around 1870. Some of Wright's engines used the Corliss steam admission controls with a more conventional slide valve

Figure 9, above. Corliss Steam Engine—Fishkill Landing Machine Company—1890s. The Corliss valves are visible at the corners of the engine block; the centrifugal double ball governor (invented by James Watt) operates Corliss's apparatus for controlling rate of steam admission; valves are activated by a connector arm extending from a cam on the main crank shaft (cam is concealed behind fly wheel) to the circular 'wrist plate' that is mounted on an axis at the center of the engine block with operating rods to individual valves. (Catalog, 7th Edition, Fishkill Landing Machine Company)

Figure 10, below. Corliss Steam Engine—Longitudinal- and Cross-Sections. In longitudinal section, cylindrical steam supply valves are at upper left and right corners, with exhaust valves at lower corners; piston is at right end of cylinder, with right supply valve open and lower exhaust valve shut; left exhaust valve is open, while supply valve above is closed (valve sequence reverses for return stroke of piston); cross-section is taken at end of piston cylinder, through center of cylindrical valves, showing shafts (for activating valve elements) extending to connecting rods at exterior of engine block. (From Uhland, Corliss Engines and Allied Steam Motors, Atlas*)*

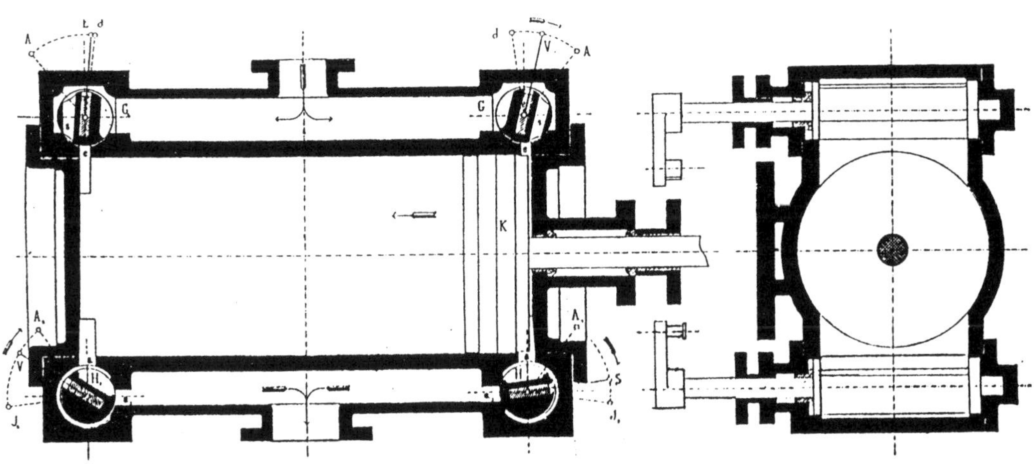

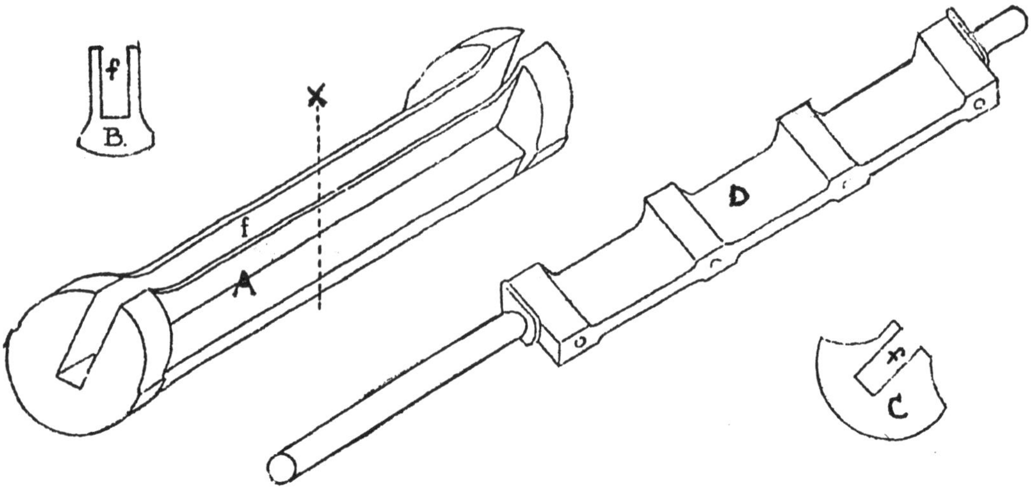

Figure 11. Corliss Steam Engine – Operable Valve Elements. Part 'A' is the entire rotating element for control of steam supply, with 'B' its cross-section; part 'C' is cross-section of the rotating element for exhaust; part 'D' is inserted into the slot 'f' in the rotating elements, with the shaft (at left end) extending outside of the engine block to a connecting rod. (From Shillito, Handbook of Corliss Steam Engines*)*

(shown in W. H. Uhland's treatise of 1879, which cited Wright's engine for excellence of design). The Bourne yard at Brockway also had a slide-valve type of Corliss engine. The Peck Company at Haverstraw , among other owners, utilized a 120-horsepower Newburgh Corliss engine, of that type.

At Fishkill Landing (now the city of Beacon's waterfront), The Fishkill Landing Machine Company manufactured a sizeable representation of Corliss engines of different horsepower (Figure 9). These were widely exported. Their engines ranged from 43 horsepower up to 780 horsepower, depending on steam pressure and piston cylinder size. The Jova Company had a 350-horsepower Fishkill Corliss. Those steam power plants had a real presence. A 100-horsepower standard 'Fishkill Corliss' engine had a fly wheel with a diameter of twelve feet and a weight of five tons! The Fishkill Landing Machine Company, founded in 1853, ceased operations in 1913 when the New York Central Railroad acquired its property to widen the railroad's right-of-way. According to the 1880 Census, there were two engines at Cordts and Hutton that had a combined rating of 128 horsepower (contradicting Sylvester's report of a single engine of equivalent horsepower). One of the Washburn yards was the only other yard in Ulster County to have more than one engine. The average total power rating at each of the Ulster County region's plants was between fifty and sixty horsepower, according to the same census. It should be noted that, simultaneous with Corliss's inventions, there was a rapid development in the water tube type of boiler design for the provision of higher steam pressures.

The Cordts and Hutton Brickyard As Model Enterprise

Sylvester, in 1880, mentions the name of a foreman, Peter Lemister, whose son and grandson also worked at the plant well past the middle of the twentieth century. Sylvester also comments on the stability of a workforce derived from local resident manpower. This stands in contrast to the importation of workers from outside of the community, predominantly newly immigrated Irish, that earlier had been required to meet rapidly expanding demands at the Downriver plants.

An early project using brick from the Cordts and Hutton yard was built in 1870. This was the school building for the Rondout community, known as Ulster Academy. After completion of the Kingston High School in 1915, Ulster Academy became a grammar school until the 1960s. The building stands today—very solidly (Figure 12)–and in 2003 is being seriously considered for conversion to residential use.

The rapid emergence of Cordts and Hutton as a principal manufacturer with a very well-equipped plant is worth a bit of conjecture. It is possible that the owners' cumulative profits from the various other business interests in which they were involved might very well have allowed them to make rapid investments in the brick business, which may have been precluded to other manufacturers without the availability of such investment funds.

Figure 12. Ulster Academy building (original name)—ca. 1870. Cordts and Hutton bricks were used. (Photo by author, 2001)

Two photographs of the Hutton Company circa 1900[8] can be used to show a considerable extent of brickmaking practice in the late-nineteenth century. Figure 13, shows a structure with a tall masonry chimney that marks the location of the boiler room and steam engines. Between the left end of this building and the portion of a roof shown at the picture's extreme left is a shallow strip of roof beneath which can be seen a thin dark line. This is the 'power line,' or continuous drive shaft, running from the centrally located steam engines to remote brick machines and tempering pits. Recall that Sylvester spoke of nine hundred feet of such power shaft, and such a shaft is shown extending across the top of the Ver Valen brick machine in Figure 4. A similar roof over such a shaft can be seen extending from the right end of the principal building in the photograph. Underneath the power line and its roof can be seen a two-wheel cart of the type that was typically used for

Figure 13. The Hutton Company—ca. 1900. Brick machine building is at rear of drying yard, with radial spokes of tempering wheels visible at each end; 'power line' common drive shaft is visible under shallow roof that projects from end of building; workers at right of central drying yard hold long-handled implements for 'spatting' the bricks; at edges of the drying yard are the vertical 'hacks' (low walls); the long, narrow, roof-shaped 'saddles' (on planks on the ground) await placement on top of the hacks. (Photo taken from kiln shed roof by Hutton Company)

the transportation of clay from the clay banks, these being visible at the top of the picture. Visible at the left corner of the building is the radial pattern of the spokes of the tempering wheel. This device also is visible under the extended roof at the building's right end. The related brick machine itself, not being really sizable, is difficult to discern. It would have to be located in back of the group of five men at the front of the building, which appears to be completely roofed with boards. The foreground is filled with the area of the drying yard, showing the patterns of brick from the six-brick molds, laid flat on the surface. This compact, geometrically precise arrangement, laid according to the 'yard line,' was used throughout the industry. The need to limit the size of the carefully constructed and maintained drying yards necessitated it. The long, thin, low walls—the hacks—are the bricks that have already been sufficiently dried on the yard's surface to allow stacking on edge, twelve bricks high. Figure 7 shows a young man (to the left of John Cordts) holding a metal implement used to turn bricks on edge (edging). The same picture shows a man to the right of John Cordts who bears a long metal implement. This is the knife used for striking off the excess clay from the mold as

Figure 14. The Hutton Company—ca. 1900. Looking north, up middle of drying yard; brick machine buildings on left, kiln sheds on right; spaces between vertically 'hacked' bricks, for air circulation, are plainly visible (bricks placed on edge); worker, at right end of first hack, holds long-handled implement for 'spatting' the bricks. (Photo by Hutton Company)

it emerged from the brick machine. Figure 14 looks north up the drying yard, showing the spaces (3/8 inch wide) between the hacked brick that allow air circulation for drying.

Both pictures show the shapes of the long, narrow 'saddles,' as the roofs placed over the hacks were named. These are lying end-to-end on the ground, awaiting placement on top of the hacks and will protect the hacks from rain. Figure 14 shows men, at the right of each picture, holding long-handled implements that have an attached flat board at the bottom. These were used to 'spat' the bricks—to remove the fins caused by the forcing of the clay into the air vent slots at the bottom of the mold. Figure 14 shows a more considerable extent of the yard's buildings, with four brick machine buildings visible on the left and the main, eight-hundred-foot-long kiln shed on the right. Between each row of hacks are the crews of men that tended each brick machine, carried the bricks to the drying yard and 'dumped' them.

Heinrich Ries's geological survey of 1895 stated that the Terry excavations had already begun to expose the rock face at the back of their clay banks, indicating that they were beginning to run out of clay, a problem that would afflict many of the yards. They eventually purchased adjacent land from the Newark

Lime and Cement Company, as did the Hutton Company. Both photographs (Figures 13 and 14) also show a feature almost uniquely unfavorable to the Hutton Company. This feature is the change of level from the machine buildings down to the level of the drying yard, a height appearing to be six to seven feet. Typically, these operations were on the same level, with the change in height necessitating, at the Hutton Company, additional manpower for handing down the molds to the yard level from the brick machine. This may be explained by the difficulty of creating a level site from the topography of the original site by the use of hand excavation, given the requirements for the large east-west dimensions of the drying yard and kiln shed. That two-level arrangement must have been both a nuisance and an expense in the nineteenth-century yard. Sylvester also spoke of the necessity of adding a forty-foot-wide strip of fill into the river for the full length of the main kiln shed—a projection that is shown on a map of 1891[9] With the means at their disposal, this filling operation must have taken a lot of doing.

How to Make a Brick

With the general appearance of a brickyard in mind, it is possible to review the entire process of turning clay into bricks at the turn of the nineteenth century. Insofar as this was the time of both the maximum number of yards, as well as the historically maximum production totals for the Hudson Valley industry, such a description will suffice to explain just how this was obtained. Turn of the century Hudson River brickmaking was a mixture of manufacturing methods that were both archaic and mechanized. This condition persisted right up to the 1920s. The process began with the excavation of clay, termed 'winning' the clay. Clay banks that were in excess of twenty-five feet in height were subdivided into terraces, or `benches,' with vertical faces of seven to nine feet in height from which the clay was dug, with hand shovels, in 'sods' approximately twelve inches square by eight inches thick. The sods were placed in two-wheel horse-drawn carts (Figures 15 and 16) where excavation was less than five hundred feet from the dump area. In excess of that distance, rail-mounted, horse-drawn dump cars were used. One manufacturer, Cary (at his Cohoes and Newton Hook yards), patented a large, movable wood covering for the clay bank—a clay house—in order to allow year-round excavation. That inventor's clay house installation is the only one known to have existed.

At certain yards in the valley, solid lumps of dried yellow clay (clay dogs) occurred. These lumps required reduction to fine material with roller mills. With dry land clay excavation eventually precluded to some of the Haverstraw manufacturers because of exhaustion of clay resources, excavation of clay from below the river's water level and dredging from the river bottom eventually became a necessity, as did the construction of an extraordinary structure like the deNoyelles's cofferdam at Haverstraw that was used to supplement their dry land clay. These excavation methods will be subsequently described.

Figure 15, above. Clay Carts—Small Ulster County Clay Bank. Note well-kept appearance of horses; more fully visible animal at center appears well-groomed and well-fed. (Photo from collection of John F. Matthews)

Figure 16, below. Clay Cart. Dumping contents at sod pit. (Courtesy of New York State Archives)

Mixing

The initial mixing, or tempering, included one of two processes. The traditional 'sod pit' was used at a large number of yards. These pits received layers of clay, sand, and coal dust (culm), which were mixed together with water and then, after soaking overnight, loaded onto steam-driven bucket conveyors for transfer to the brick machines. Also widely used was the tempering wheel ring pit for mixing clay, sand, and coal dust (Figure 17). These were steam engine-driven and operated during the second half of the day, after the brick machines had shut down.

The ring pit was twenty-five to thirty feet in diameter and three feet deep. Importantly, sufficient water was added as required (twenty-three to thirty percent of the mixture) to produce a consistency that was plastic enough to completely fill the mold space, including corners. This relatively soft consistency gave Hudson Valley brick the generic name by which they were to be known—soft mud brick. The job of proportioning materials for the mixture was of critical importance. At this point, the general division of the work day directed the events that followed mixing.

Molding

The day started early, at daybreak or before. Some workers contributed eye-openers to the early morning coffee, the brew picturesquely named 'Jump Steady' or 'White Mule.' The measure of work for the day was the 'pit,' or the amount of clay used to make the agreed quantity from each machine, ranging from twenty-two to twenty-five thousand bricks. The difference between those quantities, as the principal measure of piece work, became the most contentious subject between workers and manufacturers. The object was to complete the brick molding by mid-morning, in order to get all the newly molded brick out in the drying yard for completion of the initial drying process during the day. The six-brick molds (preferably of cherry wood, with beech and maple as suitable alternate materials), were soaked for twenty-four hours, and then dusted with sand (by the 'sander') to prevent adhesion of the clay. This sand optionally contained added mineral coloring material. The mold had to be made larger than the desired finished brick dimensions to allow for shrinkage due to water loss.

Henry Goodsell of Ponckhockie was a principal moldmaker in the Kingston area who entered business in 1880. His shop continued until 1938. Also active in Ulster County was the widely-known James B. Crowell and Son of Wallkill (founded 1872), as well as two other moldmakers in that same community. The Crowell shop is still producing molds, particularly with the resurgent popularity of molded brick at the time of this writing.

The open face of the molds, into which the clay was forced, presented corners and edges of wood. Abrasion rapidly damaged the wood during the molding process unless protected. Cast iron wear plates provided protection of these features by having cutouts in the surface of the same dimensions as the mold's com-

Figure 17. Tempering Wheel. Goldin and Ferrier yard, Catskill. (Photo by Heinrich Ries, courtesy of New York State Archives)

Figure 18. Brick Mold. Manually 'dumped' type showing protective iron facings around mold openings. (Mold from collection of Hudson River Maritime Museum; photo by author)

partments, these being screwed to the wood mold (Figure 18). The Ulster Foundry of Kingston, which was founded in the 1840s and operated to 1973, was a principal supplier of such wear plates. The brick machines of the late nineteenth century were a lineal descendant of Ver Valen's machine from the middle of the century, with a larger output and a vertical pug mill that had over thir-

Figure 19. Brick Machine Shed And Crew (late nineteenth century, location unknown). Boiler-steam engine building is at far right, with 'power line' common drive shaft visible on each side of brick machine, showing bearings on the vertical supports for the shaft; visible crew members include mold 'sander', 'molder', 'mold lander' and 'truckmen.' (Courtesy of New York State Archives)

ty knives on the central shaft, thereby providing the only true mixing that the materials received before being molded. At the bottom of the pug mill chamber were two heavy, curved 'wiper blades' (blades may be a misnomer, as they were two or three inches thick in order to push the clay sideways), attached to the central shaft, that forced the clay laterally through an opening, or die, into the press box, where the clay was then forced down into the mold by a separate plunger. The crew required for the molding operation included a 'sander' who, after dusting the mold with sand, inserted it into the machine that then mechanically fed it to the press (Figure 19). The molds were then ejected onto a table where the molder scraped off any excess clay on top of the mold. The mold was then put, by the 'mold lander,' on a flat-bedded hand truck, to be wheeled out to the drying yard by one of four 'truckmen.' The contents were carefully dropped out of the mold ('dumped') by two 'mold setters.' The completion of this work marked the finish of a half-day's work, or two 'stints,' the 'stint' unit of labor applying to one quarter of a day's work. The hand trucks for transporting the molds were equipped with springs in order to avoid excessive jarring of the newly-molded contents.

Drying

It was in the drying process where the greatest divergence in practice existed among the manufacturers, with the open yard being most widely used.

The open drying yard itself was an object of considerable care, both in preparation and maintenance (Figures 20 and 21). The need for a perfectly smooth surface that would not distort or mar the soft, newly molded brick necessitated this. The ground upon which this surface was built was given a convex shape, allowing the finished surface to shed rain water. An initial layer of gravel, eight to ten inches thick, was followed by six to eight inches of blue clay. A top layer of four inches of yellow clay was then applied. The clay was then thoroughly pulverized, rolled, graded and rolled again, "until the whole becomes as smooth as a billiard cloth"(C. E. Hall in 1905; the simile of the billiard cloth as a popular standard for smoothness being perhaps more meaningful at that time). A fine sprinkling of sand over the surface prevented adhesion by the moist brick surfaces. All such surface preparation was referred to as 'luting.' In the early spring, before commencing operations, preparation of the yard surface with large rollers drawn by leather-booted horses took place. The yard surface also received continuous care throughout the manufacturing season, and workers frequently went barefoot. The molding process had to be completed by around 10:00 AM, in order to allow sufficient time for the initial stages of drying in the open yard. After being dumped, the bricks stayed flatwise for three hours before being spatted. Spatting was the process of knocking off the clay fins that projected from the bottom corners of the bricks in the mold, because the clay was squeezed through the vent slots in the bottom of each mold compartment. The bricks were then turned on edge ('edged,' with a metal device—Figure 7) for an additional two to four hours of drying. The condition of the brick indicated to the foreman when the hacking process could safely begin.

During this entire period of time in the open drying yard, the bricks were vulnerable to damage by rain. Resulting pock-marked specimens were referred to as 'washed.' (These washed bricks enjoyed a moment of popularity in the early twentieth century.) The bricks had to be sufficiently surface-dried to allow stacking on edge and spaced for air circulation, to a height of twelve bricks—the hack.[10] The rain-shedding 'saddles' were then laid along the tops, with additional weather boarding along the sides when needed for rain protection. The bricks remained in the hacks for two to three days, depending on weather conditions, in order to evaporate most of the physically held water in the brick body. This was required both to obtain sufficient compressive strength (two hundred to three hundred pounds per square inch) for stacking into the great mass of the kilns and to prevent damage to the brick by the sudden expansion of excess water into steam during the initial firing process.

A small minority of manufacturers used other drying methods. Several yards used a saw-tooth roof system with adjustable shutters (for admission of sunlight)

Figure 20, facing page, top. Drying Yard—probably Haverstraw. Brick machines at rear edge of yard; bricks have finished second stage of drying (having been dried on edge), with building of vertical 'hacks' under way; note immaculate condition of drying yard surface. (Courtesy of New York State Archives)

Figure 21, bottom. Ulster County Brickyard – exact location unknown. Immaculately 'luted' drying yard, hacks, brick machines at rear of yard and boiler-steam engine building at center. (Photo from collection of John F. Matthews)

Figure 22, above. Pallet Yard—location unknown. Bricks are on wood pallets, supported on vertical racks with roofs over top. (Courtesy of New York State Archives)

as protection from the weather. This `covered yard' still allowed for the brick to initially be laid out on the yard surface and subsequently hacked in the same manner as in the open yard. Like other methods used by the Hudson River brickmakers, the covered yard also had English origins. That climate had an even greater requirement for sheltered drying conditions. Such drying was mostly limited to Haverstraw, with the exception of its use at Dennings Point and at the Walsh yard in Catskill. The adoption date of this system is difficult to discern (the Catskill covered yard is shown on an insurance map of 1889), as is the commencement of use of the pallet system that began with dumping the bricks onto a wood pallet (plank) immediately after exit from the brick machine. The pallets were then transported to a covered area where they were transferred to vertically arranged racks for drying. This allowed the use of much less area than was required for either the open or covered yards. This was a more critical feature in Haverstraw, where space was at a premium—and that would be the only known location of pallet yard usage with the exception of the Jova yard (north of Newburgh), which used a pallet yard for a portion of its brick drying, in addition to its open yard. The pallet yard system obviated the edging and hacking, an obvious saving in manpower, as well as greatly reducing the possibility of damage to the soft brick. Such protected drying conditions also allowed molding brick for a full day instead of a half day, as well as the operation of brick machines during rainy weather.

A Properly Dried Brick

The completion of drying, to a moisture content of no greater than five percent, was eventually realized as an optimum standard for moisture content. Attaining this degree of dryness was made more difficult because of the high water content (twenty-three to thirty percent) of the soft mud process. Pure, fine-grained clays like those in the Hudson Valley have fine pores, thereby creating resistance of the movement of water to the brick's exterior for removal by atmospheric evaporation. This also required additional time for drying, as well as creating significant shrinkage problems. The solution to this problem was the addition of sand, eventually standardized to comprise twenty percent of the mixture. This 'opened up' the clay, creating sufficient porosity to permit more rapid drying and to reduce shrinkage. It was imperative that the brick be sufficiently dry before firing in the kiln, because excessive moisture in the brick produced damaged 'cores' (interior portions of the brick) upon firing. Drying to a moisture content that was significantly lower than five percent created excessive brittleness.

The Kiln

With completion of drying, the bricks had sufficient compressive strength to allow stacking to the height required for the kiln—fifteen to sixteen feet. The surface of the kiln floor was either graded sand or surfaced with hard-burned brick and given a slightly concave shape that stabilized the outer portions of the mass by causing the stacked bricks to slightly lean toward the center of the kiln. The setters placed ('set') bricks in a pattern called 'three-over-three.' This was a universally understood system that consisted of a basic unit of three bricks set on edge and spaced 3/8 inch apart. Superimposed directly over these bricks, at ninety degrees to the long dimension of the lower bricks, were three more bricks, similarly spaced. This easily can be visualized by spacing the three middle fingers of one hand and superimposing the middle three fingers of the other hand, similarly spaced, at right angles. The clear opening between the fingers illustrates the resulting system of small spaces that ran vertically through the kiln mass, thereby allowing both heat and very necessary oxygen to flow upward. The unit of

Figure 23, facing page, top. Initial Step in Construction of a Kiln. The middle 'arch' (the 'blade'); building of kiln progressed on both sides of this central arch, toward the outer ends. (Courtesy of New York State Archives)

Figure 24, bottom. Detail of Kiln. Three-over-three arrangement of bricks, and construction of 'lower bench' (from kiln 'hearth' to top of arches); 'mudded' exterior covering of 'double coal' bricks is seen at right end of kiln. (Courtesy of New York State Archives)

Figure 25. End of Kiln; showing shoring (for stability) of kiln wall, coal piles along length of kiln and firing crew; roof boards over top of the kiln have been removed for venting of heat and steam. (Courtesy of New York State Archives)

three-over-three was repeated for the height of the kiln above the openings, and the kiln setter had to be very careful about maintaining the uniform spacing of the bricks to assure movement of oxygen-bearing air through the kiln mass. "The scove kiln is like a chimney of large cross-section with lots of baffling in it."[11]

The next larger organizational unit of the kiln was the 'arch,' a term derived from the opening at the bottom of the kiln wall that allowed introduction of fuel for the heating process.[12] These openings, spaced approximately four feet apart along the kiln wall, were symmetrically placed on each side of the kiln mass and interconnected by a tunnel that was made of the newly dried (green) bricks. The distance across the kiln—its width—was approximately thirty to forty feet. The "arch" was considered to be that four foot wide mass of bricks related to each pair of opposed openings and their connecting tunnel, for the full height and width of the kiln. Each of these segmental arch sections of the kiln contained thirty to fifty thousand brick, and the number of 'arches' in a kiln ranged from eight to forty. The number of bricks in a kiln therefore generally ranged from between four hundred thousand to two million bricks, with an average of twenty arches, or one million bricks in each kiln. Each arch virtually functioned as an independent kiln—a ticklish situation to deal with, in terms of temperature control.

'Wheelers' brought the bricks to the kiln from the drying yard on wheelbarrows and tossed them up, four at a time, to setters, who placed them in the kiln. Setters were paid as skilled workers. Bricks literally seemed to float upward, from one man to the next. The face of the kiln was stepped back at three levels, so that each man's head was approximately at the level of the next higher man's knee. Recall that the mixture included pulverized coal, or culm (in the amount of one per cent) that greatly decreased the time required for firing in the kiln. In finishing the kiln, the sides and ends were covered with a layer of brick that had a coal content of eight percent, these being called 'double-coal' bricks. The function of the coal was twofold. Upon ignition within the brick, it provided an internal heat source that insured adequate firing through the brick's thickness. It also made up for a net subtraction of heat by a reaction process (endothermic) resulting from the dissociation of the limestone content of the clay toward the higher end of the kiln's temperature range. Importantly, as explained in Chapter 1, the coal greatly reduced the amount of time required for kiln burning. With respect to the addition of coal to the mix, the leading American trade journal[13] stated that, "Brickmakers of the Hudson Valley are especially addicted to the custom." The double-coal bricks on the top, sides, and ends of the kiln had the effect of an electric blanket, in providing a hotter outer surface that prevented the escape of heat from within the kiln. Those same double-coal brick also created considerable waste (or unusable brick)—on the order of three percent of the entire kiln.

The final step in the preparation of the kiln was to cover the walls with a clay mud 'daub' that both prevented the escape of heat and infiltration of air that could both cool the bricks and reduce the required draft through the arch openings at the ground level. That mud daub was the origin of the name 'scove,' which became the generic term for this type of kiln—the kiln was 'scoved' with the mud daub.

Wood continued to be the chosen fuel for kiln burning at a large majority of the yards, right into the last decade of the nineteenth century. The Terry Brothers (at the Kingston Steep Rocks yard) changed to anthracite coal in 1884, the first use of an alternative fuel. Other manufacturers followed suit over the next ten to fifteen years. Caleb Davis Lent was also reputed to be the first user of anthracite kiln fuel at his yard in Glasco. The Terrys' yard was a logical location for such a change, insofar as the northern terminal of the anthracite-hauling D&H Canal was in the nearby Rondout Creek. The industry subsequently changed to cheaper bituminous (soft) coal kiln fuel. Heinrich Ries's study in 1900 showed the fuel costs of burning bricks (per thousand) at sixty to seventy-five cents for wood and at forty to fifty cents for coal, making a compelling case for switching from wood to coal. Oil, only beginning to enter usage, was identified as the cheapest fuel, with the only available cost information stating a royalty of $1.60 per burner (located at each arch opening). It was imperative to stop using wood, insofar as the resources were becoming very limited. It is at this point that the dried brick-to-be begins its transformation into a hard, red BRICK. Becoming a brick was a lot more of an odyssey than one might think—and a few words relating to ceramic science are absolutely essential to understand the unique problems Hudson River clays presented.

Firing a Kiln

Fuel was initially introduced on the windward side of the kiln, in order to insure adequate draft (movement of air into the kiln). It was then introduced on the leeward side after establishing draft. Initially, the heating was localized near the outer part of the kiln and then moved toward the center. This helped to obtain even heating of the entire mass, insofar as the outer surfaces had the highest rate of heat loss. The objective was to get the entire mass of unburned bricks to the same required higher temperatures at the same time. The meeting of the two lengthening lines of burning fuel from each side of the kiln, at the center, was called 'crossing the fires.' This occurred after forty to sixty hours. The sure sign of the end of the initial firing phase was the cessation of the very visible 'water smoking,' a process caused by the driving off of water between the clay particles (the hygroscopic water in the capillary spaces that coated the particles).

Excessive steam, generated from insufficiently dried bricks, could obstruct the flow of air through the kiln, and the resulting condensation on the bricks could produce damaging residues (scumming). Removal of water therefore required slow heating of the kiln mass and subsequent gradual increase of heat in order to allow complete oxidation of the brick cores and combustion of the bricks' internal coal content, at 570 to 1,020 degrees Fahrenheit (all subsequent temperatures are given in degrees Fahrenheit). Combustion of the coal required an adequate supply of atmospheric oxygen, and the coal's carbon content was removed in a gaseous form as carbon dioxide. That is why an adequate air supply through the openings created by the three over three brick pattern is essential. It is this oxidation—or combustion—of the coal's carbon content that adds significant heat to the kiln's firing process, recalling James Woods' revolutionary contribution to Hudson River brickmaking.

An oxygen-rich air supply was also necessary to prevent a change in the state of the iron oxide content of the clay, from the naturally-occurring, oxygen-rich ferric oxide to the ferrous oxide state. Ferric oxide burns red and is stable at high temperatures (it is refractory). Ferrous oxide, at relatively low temperatures, burns black and becomes molten. There follows a series of chemical changes—dissociations, or decompositions—due to increasing temperatures that result in the breakdown of the limestone that was glacially deposited in Hudson River clays. The generally recognized problematical qualities of calcium—an essential component of limestone—are attested in the remark by Hermann Seger. That great German ceramist stated that calcareous clay "does not stand in fire"—it is a real problem to fire in a kiln. It is necessary to hold the kiln's temperature in the

On Becoming a Brick

Increasing temperatures within the kiln caused a large number of significant changes in the materials comprising the brick-to-be. In a similar temperature range as the internal coal combustion was the separation of the chemically combined (hydroxyl) water from the layered silicate (clay substance) particle, this occurring from 840 to 1110 degrees F. Dissociation of the limestone begins around 1,450 degrees F., liberating its carbon content (from its natural condition as calcium carbonate, $CaCO_3$) as gaseous carbon dioxide, transforming the calcium into the highly reactive calcium oxide form (CaO) that becomes the nemesis of the region's clay.

Naturally-occurring clay that is uncontaminated with calcium begins vitrification (fusion of clay particles) between 2,070 and 2,290 degrees F., depending on the purity of the clay, with the spread of temperatures (the firing range) between the beginning of vitrification and complete vitrification varying between 90 and 180 degrees F. This is a workable spread of temperatures, allowing control over the gradual fusion of the clay particles and producing the desired strength and reduction of water absorption as required. Even in such clays, there are elements present that lower the vitrification temperature relative to the purest type of clay substance (Kaolin), which vitrifies at 3,200 degrees F. Enter the villain calcium which narrows the firing range to 35 to 55 degrees F. in Hudson River clays—impossibly close, given the difficulty of temperature control in the scove kiln—and very difficult with any other type of kiln. Vitrification is abrupt, beginning at around 1,825 degrees F. With complete vitrification, the brick becomes a solid, nonporous mass, shrinking by at least fifteen percent (the pore volume of the unvitrified brick). Total liquification follows complete vitrification after an additional 50 to 60 degrees F. temperature increase. This results in a total loss of the kiln contents, and the whole thing becomes a sea of molten glass. Such liquification could also occur in very localized portions of the kiln, adjacent to the firing arches. All of this precluded moving above the 1,800 degrees F. temperature that begins clay particle vitrification, with a workable temperature range for adequate maturing being limited to 1,700 to 1,800 degrees F. (1,600 degrees F. as absolute minimum). How else to glue it all together? That is accomplished by the new compounds that are created after the decomposition of the limestone and the fluxing of the calcium oxide. These new compounds are molten above 1600 degrees F, flowing around the unvitrified clay particles, and are termed 'fluxes' (from Latin *fluxus*—flow). In this case, the calcium oxide reacts with the clay substance elements of alumina (Al_2O_3) and silica (SiO_2) to form stable types of silicates, closely resembling clay substance, that occur in nature—namely anorthite and diopside that both contain calcium. Fluxing at 1,600 degrees F., these compounds provide good hardening and little shrinkage—and considerable porosity because they do not fill the voids between the clay particles and sand. That porosity became a source of intense debate and concern in future years. Such bricks are an atypical ceramic product, in that the clay particles are not actually fused, but used in a similar manner as the finer aggregates in concrete—stone (of graded sizes) and sand (smallest size)—that are glued together by cement. In bricks, the sand is the largest 'aggregate' size, with the clay particles filling in between, all glued together with anorthite as the most commonly occurring flux.

1,700 to1,800 degrees range for ten hours—in order to 'ripen,' `soak,' or 'mature' (commonly used terms for this part of the process)—so that a range of chemical dissociations and recombinations into stable new compounds can take place.

The bricks in the arch tunnels went through a very visible change in color, which was a workable indicator of temperature to the experienced eye. At 1,252 degrees, the bricks glowed dark red, at 1,666 degrees cherry red, and at 1,832 degrees bright red, cheerily and ominously signaling that the kiln was close to disaster if allowed to get to this last coloration. A skilled fireman could make sure that the fiery scene in all of the tunnels looked the same.

Bricks that have been properly hardened in the kiln will give the nice, sharp 'clink' sound, when struck together, music to a brickmaker's ears. Even with the ideal finishing (soak, or maturing) temperature at 1,700 to 1,800 degrees, the problems in simultaneously raising all parts of that mass of bricks to that optimum temperature were too great to avoid consequential amounts of unusable bricks—as high as ten to fifteen percent in the late nineteenth century and significantly lowered to six to seven percent in the mid-twentieth century. The underburned bricks were classed as 'pale.' Those closest to the kiln arches, subject to the sudden introduction of heat and the highest temperatures where they would swell, rupture, and blacken, were termed 'lammies' (also called clinkers). A mid-twentieth-century authority (Amberg) stated: "Every kiln is an adventure with one million to one and a half million brick. Every kiln that is opened is full of surprises, pleasant and unpleasant. So many factors can influence the burning of the scove kiln that the fireman never knows what to expect and must be ready to change his procedure to meet developing situations." Hence the need for conscientious kiln foremen who took this critical job very seriously—and they were at the top of the pay scale, along with the barge captains.

In a manufacturing technology requiring considerable control of temperatures, it must be said that the scove kiln presented unique challenges. Upon completion of firing, the entry of air through the kiln arches was blocked with iron covers across the arch openings, in order to prevent excessively rapid cooling of the kiln mass. The time required for burning the older scove kilns was one week, and was approximately the same amount of time required for cooling. With the later advent of oil as fuel and forced-air draft, the burning time was reduced to four to five days.

The Hudson River clays' calcium content also lowers the bricks' dried strength by somewhat more than twenty per cent, as well as lowering the fired strength of the finished product by almost ten percent—all in comparison with the same clay without the calcium contamination.

Moving Finished Bricks to Market

Until mechanization of kiln sheds, the kilns were dismantled in the same manner in which they had been built—by being manually handed down in stacks of six bricks (two more than when they were handed up). They were then placed on wheelbarrows, wheeled out to the schooner or barge, and hand-stacked on the deck (also stowed below in the hold of schooners). Upon arriving at a berth in the city, they were again removed by hand from the deck to the dock. The bricks had been manually handled by individual men, in small stacks or as individual bricks, for a total of twelve times at the yard—plus the off-loading at the city dock. Needless to say, this kind of labor-intensive materials handling became the subject of intensive efforts and ingenuity for removal by mechanization.

New and Alternative Technologies

The last quarter of the nineteenth century saw the introduction of new technologies that were adopted in a limited way by the Hudson River industry. A brief survey of radical, new technologies available to the American brick industry during the nineteenth century is necessary, in order to understand the limitations placed upon the Hudson River manufacturers by the available clay resources and by the market for their bricks. The first of these represented the abandonment of the ancient process of forming the brick with a mold. In this radically new process, the clay was forced—extruded—through an opening, or die—like a squib of toothpaste (Figure 26). This length of extruded clay, having the dimensions of the length and width of a brick, was then cut to required thickness dimensions. (The thickness dimension is actually the height of a brick when seen in a wall.) Alternatively, the extrusion could have the height and width of a brick and be cut to the required length. The origins of this process are difficult to discern. The earliest evidence appears in England during the 1830s—possibly a plunger type that exerted a pushing force on the clay that had the disadvantage of being intermittent, due to the reciprocal nature of the mechanism. In 1853, James Penfield designed and built an American plunger type of extrusion machine. Penfield's name was coincidental with another clay machinery inventor of little more than a half century later.

Figure 26. Mouthpiece of Extruded Brick Machine. (From Dobson, A Rudimentary Treatise on the Manufacture of Bricks and Tiles*)*

The development of the continuous auger-fed extrusion process (like a sausage-maker) overcame the shortcomings of the reciprocal machine. Two such machines were introduced in the 1850s. In the United States, machinerymaker Cyrus Chambers of Philadelphia experimented with such a machine during the late 1850s, with first actual usage in 1862. Another auger-driven extrusion machine was introduced in Breslau, Germany, in 1858. One authority (the journal *Brick*) considers this to be the initial appearance of the auger design.

The extruded bar of clay was cut to the thickness (or length) of the brick by wires contained within a frame—similar to a hard-boiled egg slicer. The use of the extrusion process spread rapidly throughout the world of brickmaking, with the

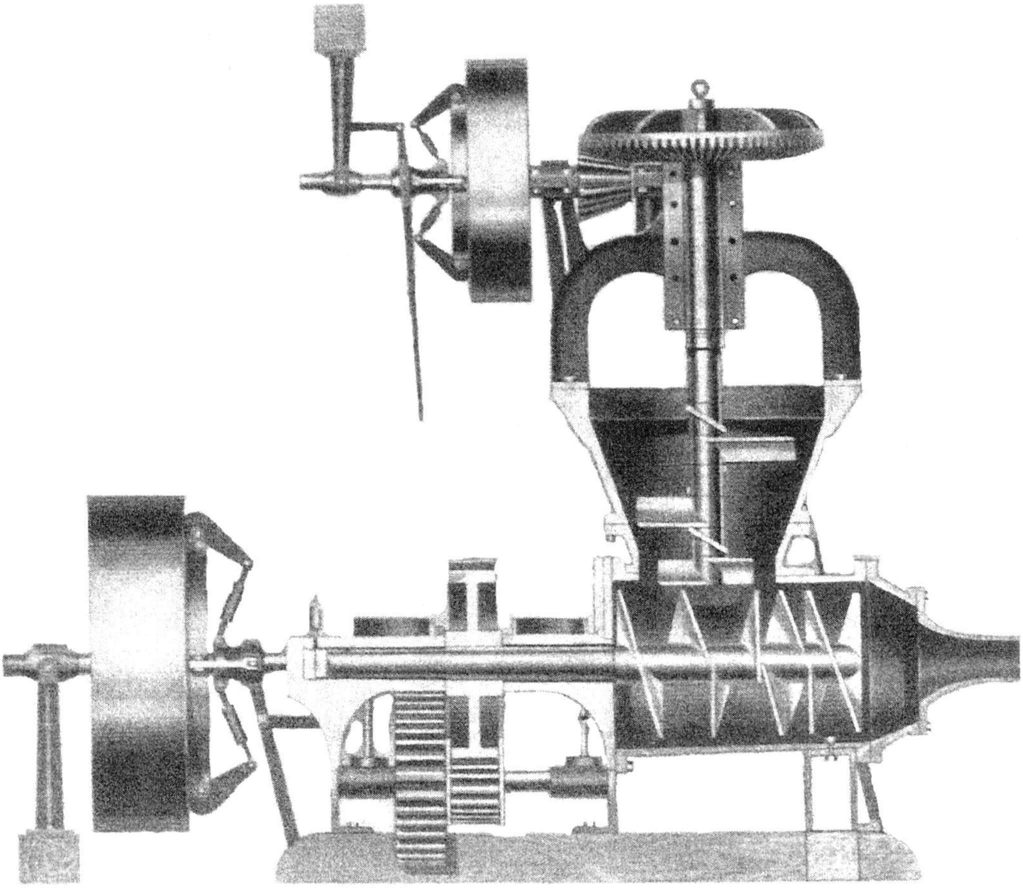

Figure 27. Longitudinal Section of Auger-type Extruded Brick Machine. Horizontal auger is at right end of machine, with extrusion mouthpiece at extreme right; this machine has a vertical pug mill feeding into the auger chamber; belt-driven power takeoff wheels are at extreme left of auger shaft and at top middle, for driving pug mill. It had a capacity of seventy-five to one hundred thousand per ten hours and weighed twelve thousand pounds. (From advertisement by Marion Machine & Foundry Co. in Brick and Clay Record*)*

exception of the Hudson River industry (Figures 27 and 28). Significant advantages of this system were in the simplicity of the extrusion machinery, its potential for large output, and its use of significantly less water in the mixture—thirteen to eighteen percent for the extrusion process versus twenty-five to thirty percent for soft mud brick. In comparing the operation of extrusion and molding machines, an early-twentieth-century analyst commented on: "the working of the American soft mud machine, which is rather expensive." The difference in moisture content identified extruded brickmaking as the 'stiff mud' process. The extrusion process also permits perforation of the brick with holes that decrease weight and allows a more even exposure to kiln heat throughout its thickness. In addition, it allowed the production of much larger bricks than was allowed by the limitations of soft mud molded brick production.

The absence of the adoption of the stiff mud process by the Hudson River manufacturers was positively not the result of provincial backwardness. Recall that the local industry had already made its own considerable contributions to brick manufacturing technology. Actually, a history of Dutchess County[14] states that a Cyrus Chambers' extrusion machine was to be installed at a Fishkill Landing yard. The outcome of this particular experience is not known, nor whether it revealed the shortcomings of the extrusion process with the local clay. At any rate, the extrusion process was certainly known, with at least one very interested manufacturer reportedly involved in its use.

This introduces the second fatal flaw in the Hudson River clay material itself, that flaw being its small particle size and consequent high 'plasticity.' Such clays have an excessive tendency, during an extrusion process, to develop planes of weakness termed 'laminations.' The defects show up within the thickness of the bricks after firing and are an unacceptable flaw. A principal cause is in the frictional resistance of the die opening's surfaces to the flow of the clay, thereby creating a lower velocity at the perimeter than toward the center of the extrusion. Another cause is the separation of the advancing column of clay into two parts by the auger, creating a plane of weakness when forced back together (into a single column of clay) during the actual extrusion process. There is also a tendency of the metal surfaces pressing on the clay to move moisture away from those surfaces, thereby creating problematical moisture imbalances across the clay column. Heinrich Ries, in 1900, stated, "that the [Hudson River] clay does not stand treatment by other methods. Stiff mud machines seem to be barred out completely by

Figure 28. Auger-type extruded Brick Machine. the cutting table is at left; extrusion mouthpiece is at center; clay (or ground shale) was introduced into top of machine at right; belt-driven power takeoff wheel is seen at extreme right. (From Dobson, A Rudimentary Treatise on the Manufacture of Bricks and Tiles*)*

the nature of the material. . . ." This was confirmed in 1951 in a report by the very competently led cooperative laboratory established by the Hudson River manufacturers.

This eventually fatal flaw did not become troublesome until the 1950s. Acknowledgement of other limitations on alternative uses of the river's extensive clay deposits should also be made, recalling Heinrich Ries's critical assessment of 1900: "Many attempts have been made to utilize the vast deposits of clay found in the Hudson Valley for other purposes than common brick, but thus far only cases of failure are recorded." Circumstantial evidence alone provides ample information to prove this, in a dearth of ceramics industries other than brick manufacturing. This is in contrast to Ohio with its enormous ceramics industry that included not only significant brick manufacturing, but also very large dinnerware (whiteware) manufacturing at East Liverpool (hundreds of manufacturers), electric insulators, sewer tile, and art pottery (Rookwood, Zanesville), etc.

One other method of forming brick—the dry clay method—seems to have had very limited application in the American industry. This method was largely limited to the production of a highly finished brick with complex, molded decorative surfaces for which there was a very limited demand.

The Hoffmann Kiln

The second radical innovation of the nineteenth century was the 'continuous kiln,' thereby introducing the terminology for two fundamental types of kilns—'intermittent' (or periodic) and 'continuous.' The scove kiln is of the intermittent type, being built, fired, cooled, and dismantled. The continuous kiln is continually in operation. In 1858, Friedrich Hoffmann, the German State Architect, invented the latter type. It was patented in the United States in1865. His kiln's mode of operation reduced fuel consumption by sixty percent of that required for existing types of kilns, through the reclamation of heat from newly fired (cooling) bricks, and by the use of heat from actively fired portions of the kiln for pre-heating unfired bricks.

In 1912, journals show that continuous kilns' fuel savings were on the order of forty percent—a disparity with the earlier figure. Hoffmann received considerable assistance in the development and promotion of his kiln from the great German ceramic scientist, Hermann Seger (distinguished for work in both industrial ceramics and fine arts porcelain development), whose writings were translated and published in the principal American trade journal *Brick*. Seger, in writing about the Hoffmann kiln, noted the problems presented by lime-bearing clays and promoted the use of gas as a kiln fuel (undoubtedly 'producer gas' derived from coal).

Use of this type of kiln spread rapidly through various parts of the world of brick making—except along the Hudson River. Such a design consisted of a series of chambers, initially arranged in a circle, and then in a flattened oval shape requiring a substantial, permanent masonry kiln structure (Figures 29 and 30).

While firing the bricks in one chamber, the heat from that chamber was introduced into the adjacent three chambers, full of unfired, dried bricks, thereby using that 'borrowed' heat to substantially raise the temperature of the unfired bricks. Additional fuel provided to successive chambers completed the firing. Where required, chambers were commonly separated by paper dampers that were removed at relatively low temperatures (circa 240 degrees). Such dampers were only necessary to control air flows in chambers that had relatively low temperatures. In addition, fresh air for fuel combustion was drawn past the freshly fired (cooling) bricks and into the actively firing chamber, thereby heating the combustion air to high temperatures, at no additional cost. The process continued around the circuit of chambers. Because of the movement of air through the chambers, these kilns were also known as 'horizontal draft' kilns. Those processes of heat reclamation were the genius of the Hoffmann kiln. In 1900, each chamber typically contained twenty to twenty-two thousand bricks. That number greatly increased by the end of the decade.

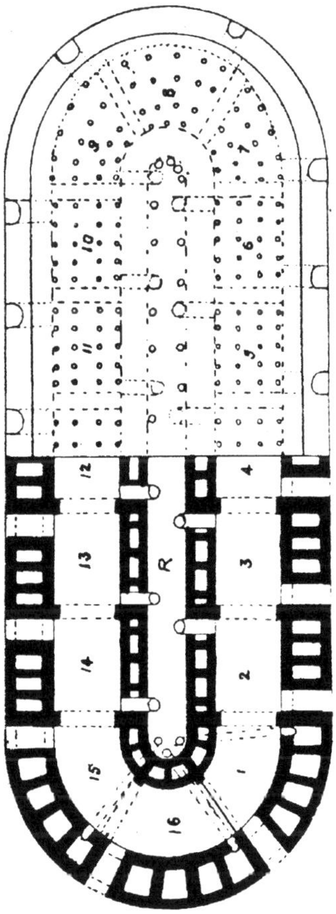

Figure 29. Plan of Hoffmann Kiln. The upper half of the diagram shows the roof of the kiln with firing ports (small circles); the lower half is plan at floor level, showing circuit of chambers and double wall construction; the narrow space in the middle ('R') relates to the kiln's flue system, to which each chamber was attached. (From Dobson, A Rudimentary Treatise on the Manufacture of Bricks and Tiles*)*

Hoffmann's original design was for a circuit of twelve chambers, with this number eventually perceived as impracticably low. In the kiln's final form, a total of sixteen to eighteen chambers was required for the complete cycle of filling with unfired brick, pre-heating, firing, cooling, and emptying, at a rate of one chamber actively fired per day, or six to twelve inches per hour along the length of the kiln.[15] Multiples of sixteen to eighteen chambers could be arranged in a single loop, with the world's largest (American) at over sixty chambers. This type of kiln, with its ducted exhaust system for combustion gases, also allowed for the option of heat reclamation for drying newly formed bricks either before or after setting in the kiln. The Hoffmann kiln also had very good firing characteristics, due to the relatively even temperatures applied to a limited number of bricks within each chamber, thereby producing high percentages of usable bricks. Heinrich Ries's writings of 1895 make no mention of Hoffmann kilns. His work in 1900 on the clays of New York provides the first evidence of general availability, noting the first usage during the last five years of the 1890s.

Figure 30. Hoffmann-type Continuous Kiln. Note loading doors and elevated platforms containing coal fuel for firing at top of kiln; also visible at the loading doors is the battered (slanted) wall construction that was required to counteract the horizontal thrust of the masonry roof structure. (Courtesy of New York State Archives)

Another Intermittent Kiln

The 'downdraft' or 'beehive' kiln was the second type of intermittent kiln to appear during the late nineteenth century. Scove kilns, with their rising heat, were 'updraft' kilns. Beehive kilns were dome-shaped, permanent kiln structures of approximately thirty feet in diameter (Figures 31 and 32). They had firing ports around the perimeter for fuel introduction. Heat rose to the top and then was directed downward through the bricks to outlet flues in the bottom of the kiln. Downdraft kilns produced very even temperatures and a high percentage of usable bricks, but were never capable of high production volumes. They were also more expensive to use in that their structure required constant attention and maintenance. Excessive heat loss through the kiln walls, which was exacerbated by a much larger ratio of exterior kiln surface-to-volume within the kiln, also contributed to high fuel costs. Deficiencies in flue design also plagued a significant number of downdraft kiln installations. This was very critical due to the need for a free flow of hot air that was traveling downward for part of its movement through the kiln, followed by horizontal movement through underground flues to a high vertical exhaust stack. Scove kilns were estimated to use thirty percent less fuel than the downdraft type. Like the scove kiln, beehive kilns were of the intermittent type in that they were loaded, fired, cooled, and unloaded. Such kilns were widely used in New England.

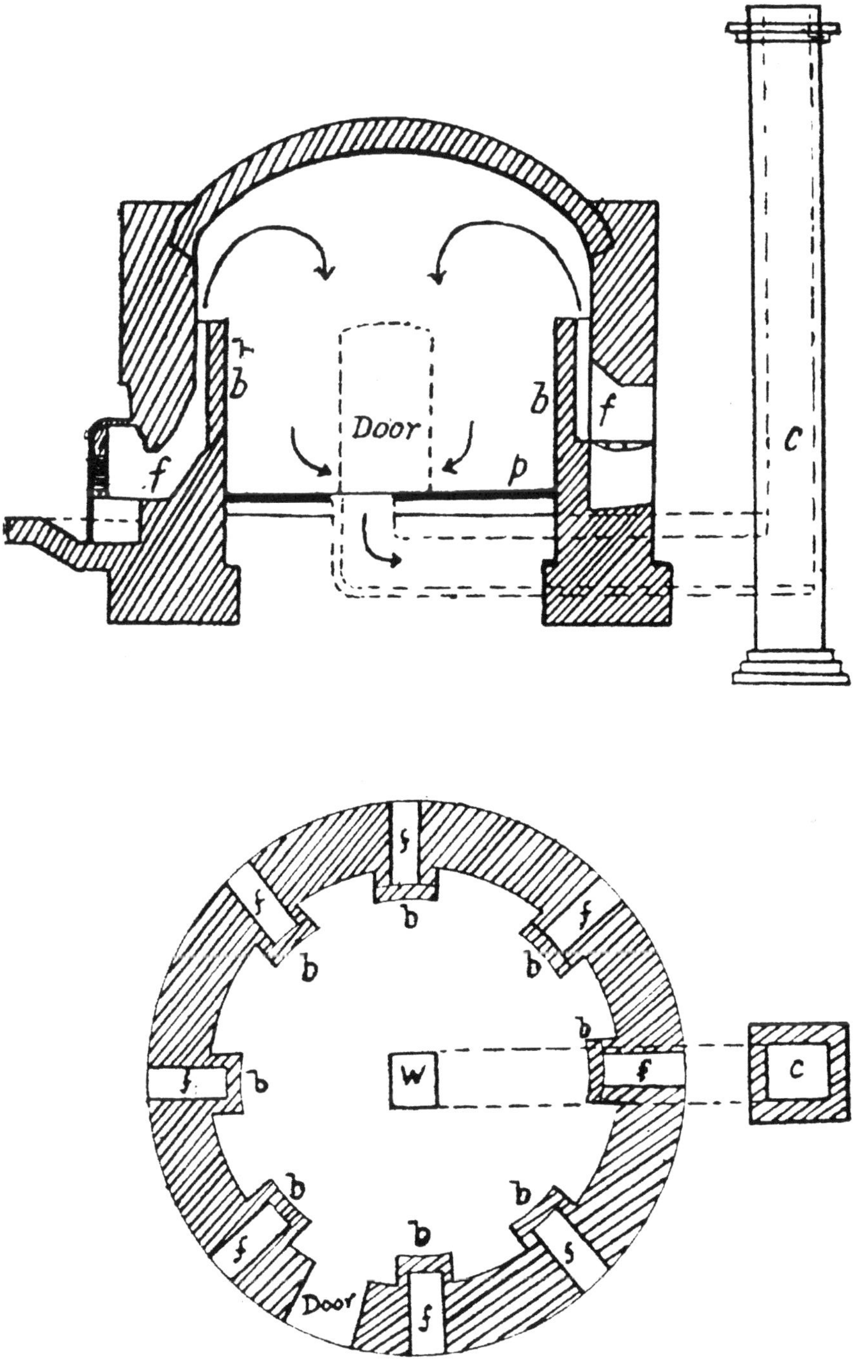

Figure 31. Beehive Kiln (Downdraft Kiln Type)–Vertical Section and Plan. Note vertical hot air movement from firing ports to top of kiln space, with downward movement to underground, horizontal flue leading to vertical stack. (From Dobson, A Rudimentary Treatise on the Manufacture of Bricks and Tiles*)*

Figure 32. Beehive Kilns. Steel bands are for stabilization of the structures. Flues for each kiln are visible at rear. (From Brick and Clay Record, *1912)*

Progress Versus the Scove Kiln

Regarding the Hudson River industry, the authoritative Heinrich Ries raised a very vexing issue in 1900: "why the old out-of-date scove kilns still hold their own is a matter hard to explain." A great deal will be made of this criticism within this writing. Ries also seemed to contradict himself, or at least to answer his own question in another statement about the merits of scove kilns. The Hudson River market demanded enormous volumes of a range of brick grades, including the use of 'light hards' for partitions and non-bearing backup, hard-burned 'prime' bricks for load-bearing structural conditions, and flawless, hard specimens for use on exterior surfaces. These were all supplied in the general production run of late nineteenth- and early-twentieth-century Hudson River common bricks. The rapid building of the repetitive series of scove kiln arches, with their simultaneous firing, was capable of providing enormous quantities of low cost bricks. Detailed description of contemporary kiln performance is almost nonexistent, with an exception in an article from the *Clay Record* (March 1908) that stated a burning time requirement for both continuous and downdraft kilns at eight to ten days, with an equivalent cooling time—for relatively small volumes of bricks. (That production rate concurs with Searle's previously stated rate requiring sixteen to eighteen days for completion of a continuous kiln circuit.) This was in marked contrast to the ten to twelve days required (firing and cooling) for the much larger scove kiln—a sizable consideration that counted against the continuous and downdraft kilns insofar as they were not equally high volume producers.[16]

In spite of his criticism of the scove kiln usage, Ries went on to list significant advantages of the scove kiln, including economies of construction, maintenance, and operation, with the latter feature of particular note—and, most notable of all, their large capacity. He also directed attention to the higher percentage of unusable bricks—and this was certainly the principal negative for the scove kiln. Neither the Hoffmann type of continuous kiln nor the beehive type of intermittent kiln ever appeared in the Hudson River common brickmaking world, with one notably unique exception in the case of both of those kilns. Was this poor judgment, as implied by Ries, or was it a calculated assessment for meeting the demands of a market that, from 1890 to World War I, had made unremitting demands for enormous quantities of low cost common brick product?

Among that large number of producers were some very capable and informed men who certainly had familiarity with all the available technology choices—and were also capable of ascertaining the benefit of such choices to their own circumstances and the risks involved in the commitment of capital to costly new methods. Those gentlemen appear to have deliberately avoided the use of the alternative kilns. Innovation was quite familiar to the manufacturers, insofar as the Hudson River brick industry had already compiled an enviable record of inventiveness over a period of seven decades, and there was more to come. The Hoffmann and beehive types of kilns could not provide the enormous quantities of low-cost bricks that the market demanded. Final answers to Ries's criticism will be provided in Chapter 3 and in Chapter 6.

Other Evolving Technologies

Dryer technology was also evolving. The same American machinery manufacturer (Cyrus Chambers of Philadelphia) that had devised a stiff mud machine did, in the 1860s, introduce a radically different procedure for drying. This consisted of a forty-foot-long tunnel in which rail-mounted cars bearing newly-formed bricks received warm air or radiant heat for drying. This was accomplished in eight to twelve hours. One account states that William Van Vliet's yard on the tidewater portion of the Fishkill Creek used hot air drying during the late-1850s. Unfortunately, this yard was too distant from deepwater access, causing its failure in the absence of steam-powered towing. Variations on tunnel dryer design developed through the rest of the nineteenth century, with such a dryer in use at Haverstraw's Diamond yard by 1900. This method may very well also have had difficulties drying the very large production of the Hudson Valley yards. In the early twentieth century, local manufacturers adopted another type of artificial drying environment.

Explosive Urban Growth and the Need for Bricks

The demands of the metropolitan market have already frequently been mentioned. The explosive growth of New York City, both in commerce and in population, drove this ravenous demand for brick. Recall that the fire limit had been set at 32nd Street in 1847. By the end of the nineteenth century, dense development could be seen north of 125th Street. Development had rapidly expanded into Harlem by means of rail transport as early as 1837. An elevated railroad supplied real mass transit after 1880. Harlem was largely built of brick. In the 1860s, a moderate foreign immigration accompanied an outflow of 'native' population. The three decades following 1870 showed a very large increase in native population and an increasing rate of foreign migration. The 1870 population was 1.3 million. By 1890, this increased by exactly one million. A staggering increase of almost 2.5 million human beings occurred in the next two decades and 1.15 million were added in the 1890s decade alone, for an increase of forty-eight percent! This was the period of the great migrations from the Old World, resulting from transatlantic passage suddenly made cheap by a large number of steamships, and because of economic and social calamities in Europe. Where to house all these new arrivals?

By 1900, 82,650 tenements housed seventy percent of the city's population. The initial responses to the great need for housing for this massive influx of people, by many of the developers, was severely deficient because of scandalously inadequate regulations. Post-Civil War housing regulations, beginning with the Tenement Housing Act of 1867 (a rudimentary regulation at best), attempted to deal with reduction of lot coverage by the building, better ventilation, access for light, sanitation, and enforcement measures. It was only through the testy process of interaction between the housing reform movements (spurred on by that greatest of all reform advocates, Jacob Riis) and the political process that effective regulations could be created and enforced. Even the notorious 'dumbbell' flat—actually, the winner of a competition—was the result of a late 1870s reform effort. The stipulated dimensions of the lot for which it was designed (twenty-five feet by one hundred feet), as well as virtually nonexistent setbacks from property lines, a condition created by the limitations of the utterly inadequate lot size, were severe flaws in this reform effort. The dumbbell flat replaced the even more inhuman 'railroad flat' with its unventilated interior rooms and totally deficient amounts of space, light, and fresh air for the inhabitants.

The Greater New York Charter of 1897 brought on the new building code of 1899, resulting in the New Tenement Act of 1901. The latter regulation provided acceptable living conditions. Lot coverage was to be no greater than seventy percent for 'interior' lots (between corner lots) and had minimum rear yard dimensions of twenty-four feet (allowing forty-eight feet between rear building walls). Good bricks were indiscriminate in their usage—they could be used for buildings that could be very good or very bad. Even the bad buildings, most importantly, were fire resistant. An enormous amount of commercial construction was also using vast quantities of bricks. All of this implied a seller's market for bricks—and that was a simplistic conclusion.

Troublesome Elements of an Unpredictable Market

Other powerful influences superimposed on the increased demand for building materials, including national economic cycles such as the national financial crisis of 1873, plunged the country into a harsh recession lasting or several years. The 1870 price of bricks—always given in the cost per one thousand bricks—was $8.25, a very profitable figure. By 1879, it had plummeted to $4.08 per thousand. The break-even price was approximately five dollars per thousand. Bear in mind the extraordinary deflationary period of the time as witnessed by the Federal Reserve Bank of New York's cost-of-living index which showed a twelve-percent decrease between 1870 and 1900. Brick prices snapped back to $8.28 per thousand in 1883. Prices were thereafter subjected to continuous downward pressure, hitting a low of $4.25 in 1900, at the end of the deep nationwide depression that lasted for most of the 1890s. The price per thousand did not reach eight dollars again until 1905, twenty-two years after hitting that level in 1883. This was a very rough ride for the brick business and a challenge for any businessman who attempted to analyze risks for paying off large investments in his business. Prices hovering close to the manufacturers' production costs for over a decade, with dips below the break-even level, are not the stuff of a businessman's dreams.

There were other forces that had powerful effects on this market. One of the most significant was that a brick manufacturer did not need to own the land upon which his clay extraction and manufacturing occurred. Recall that James Wood, the progenitor of the Hudson Valley brick industry, had leased deNoyelles's land in 1815. Extensive leasing was so well established in Haverstraw by 1880 that, on the lands of seventeen property owners, there were twenty-six leases for brick manufacturing. Sixteen of those seventeen landowners also were manufacturers. Daniel deNoyelles places the number of clay landowners-leasors at ten, significantly at odds with the larger total above.[17] The number of lessees eventually went much higher. James Eckerson's land had six lessees, while Washburn, Worall, and Palmers' lands had eight lessees and deNoyelles and Company's lands had four. Some of the lessees were medium-size producers, including such yards of the Lynch Brothers and the Malley Estate, both on the deNoyelles lands—which included three quarters of a mile of waterfront. A good number were short-term partnerships of men of limited resources who pooled their assets to lease, on the average, three brick machines, as well as the clay resources for brickmaking and drying yard and kiln shed space. The lessors had capacity to spare. The average had at least seven to eight machines. Such leased operations averaged forty to sixty employees and could be in and out of business in a relatively short period of time, with the lessee making six to seven million brick. The leases were an echo of James Woods's remark from over fifty years earlier: "Brickmaking is a poor man's game." Costwise the lessors couldn't lose, because the lessee fully assumed the risks of the market. Initially, the lease rate was three thousand to four thousand dollars per brick machine. The lessee anted up this sum before commencement of brickmaking. The lessee then got a bank loan for payment of wages.

With the sudden weakening of the market in the 1890s, the lease arrangement changed to a percentage charge against gross sales, or a fixed charge, ranging from twenty-five cents to one dollar on every thousand bricks made by the lessee. The circumstance of a considerable number of customarily leased brick machines standing idle in 1894 created this new arrangement. Respecting the variable market price, the fixed lease rate per thousand manufactured was inflexible, with the previous rate per machine being even more so. The percent charge on gross sales at least made some recognition of the vicissitudes of market prices. "After each losing year, there were always new adventurers to lease clay banks and molding yards, collect a few thousand dollars with a few close friends and bargain with the long-standing landowners for a fresh go in the hazardous brick business."[18] Manufacturers who relied on leased land and equipment did not only have their own brick brand but applied that same brand to their bricks at whatever place they were manufactured. Proof of the enormous activity in leasing is the sheer number of brick brands from Haverstraw—over one hundred and fifty. Ultimately at risk were the clay resources that large numbers of manufacturers were ravaging. The Malleys quit the last Haverstraw percentage type of lease in 1927, at a rate of twelve and one half percent of their gross.

Upriver vs. Downriver

Upriver, there was a significant increase in the number of manufacturers, from eighteen in 1860, to twenty-five in 1870, and to thirty-six in 1880. The increase in the value of Upriver production was just as notable, moving from $176,000 in 1860 to $1,090,000 (up five hundred percent)[19] in 1870, with this value surpassing Rockland County's $937,800 in the same year. The plummeting prices of the 1870s dropped the Upriver production value back to $649,000 in the 1880 census[20] and Rockland County down to $637,440. The future dominance of Upriver brick manufacturers was approaching due to their access to much larger properties and clay resources. A large reduction in the number of Westchester manufacturers during the 1870s also accelerated Upriver dominance. During this time the numbers of Westchester manufacturers dropped from twenty-nine to a total of eleven, and they also experienced an eighty percent drop in the value of production. The effect of a large amount of manufacturing on leased premises added to the production capacity, with most of this limited to the Haverstraw district. The net result of the rapidly increasing production capacity was to create all the conditions necessary for an excessive supply of bricks to the market, with all the consequences to prices resulting from a buyer's market.

By 1905, the total number of manufacturers was fifty percent greater than in 1880, and the total value of production increased by 350 percent. Notwithstanding the dismal prices of the 1870s, Hudson River manufacturers nevertheless, by 1880, made the investments in steam power necessary to completely change a fundamental operating function of their yards. The reasons for doing so must have been obvious enough.

The Cordts and Hutton yard was fortunate in having access to capital that was generated outside of the brick manufacturing operations, thereby assisting them in becoming a leading brickmaker. This is admittedly a speculative proposition, but made believable by their unique position of having various resources available to them. Adequate capital and smart management were essential for survival. Such speculation introduces an opportunity to remark on the profitability of the industry to the individual entrepreneur. There were no big fortunes made in the brick business. Those who accumulated larger-than-average amounts of money seem to have done so by having substantial additional sources of capital formation, in addition to ownership in the brick industry. By itself, brickmaking was no way to get rich.

New Threats to the Brick Industry

The changing nature of construction materials usage in the metropolitan market also contributed to fluid business conditions. Favoring the brick market was an extraordinary post-Civil War relapse in the progression of metal framing development. The efficiencies of the cast iron building frame, so vigorously pursued before the Civil War in New York City by James Bogardus (a prodigious inventor in multiple fields of manufacture), Daniel Badger, Abram Hewitt, and Peter

Cooper, were not effectively advanced during the immediate postwar years. The institution (Cooper Union or Cooper Institute) that was named after its sponsor Peter Cooper, with its initial phase of construction completed in 1859, employed arched brick infills between its iron joists, with a concrete fill over the top that provided a level surface for the floor above. Ironically, the structural iron inventions of Bogardus, born and raised in the Hudson River city of Catskill (with its own very active brick industry), was ultimately disastrous for the Hudson River brickmakers. Metal framing efficiencies were to be more rapidly realized in Chicago during the post-Civil War years. Even the big Equitable Life Assurance Company building of 1870 showed equivocation. Originally designed as an all-masonry building, the very accomplished architect George Post took over the Equitable project, redesigning it as a hybrid metal and masonry building with a height of 130 feet. It made much use of brick in partitions and in the arched floor system that spanned between wrought iron joists, as well as brick infills that provided wind bracing between vertical supports.

Slowing the progress of the adoption of all-metal framing was a real concern for the safety of metal framing when subjected to high heat in fires. Even Post's mid-1870s Western Union building, almost twice as high as the Equitable building, still used masonry support for iron beams because of concerns for fire safety. A very large amount of this type of hybrid metal and masonry commercial building construction was accomplished during the second half of the nineteenth century.

Ten years after the Western Union building, Post completed a job that was almost fully framed in metal and supported on iron columns. More commonly, thick brick masonry walls, supported on iron beams, continued to be used in walls along property lines, with this type of construction also occurring across the façade. Brick walls continued to be used as wind bracing until Bruce Price's three-hundred-foot-high American Surety building of 1895. Such height precluded use of extensive masonry. Tile cladding systems, newly designed for the purpose, provided fireproofing. Why not diversify into the tile business in the Hudson River clay lands? Recall Heinrich Ries's statement from 1900: "Many attempts have been made to utilize the vast deposits of clay found in the Hudson Valley for other purposes than common brick, but thus far only cases of failure are recorded." The tallest commercial buildings no longer used the large quantities of brick that had previously been consumed in their construction. During the two decades after 1890, the enormous surge in population offset reductions in brick usage due to increasing use of steel framing. This is because the population surge required great amounts of housing construction. Popular taste, ever a fickle determinant of consumer preference, removed common brick from the nineteenth-century list of fashionably acceptable finishing materials, and even the more highly finished face bricks had significantly less acceptance in New York City than in other regions of the United States. No hard feelings—even brownstone was in trouble, according to a taste arbiter in the early 1870s: "A block of houses built of

Ohio stone, a material fast coming into use and destined to supplant brownstone. . .vastly more attractive than houses of ten years ago."[21] More important, a new and far more damaging competitor was gradually emerging, namely concrete. The development of Portland cement expedited this, with its greatly increased speed of setting (hardening) in comparison with the existing 'natural' types of cement that had been used in great quantity during the nineteenth century. A major center for production of the latter type of cement was at Rosendale, twelve miles southwest of Kingston. The encroachment of concrete made with Portland cement rapidly accelerated during the early years of the twentieth century.

Throughout the foregoing period of time, an enormous number of smaller buildings were being built that required large quantities of brick masonry. A four-story exterior wall required twelve-inch thickness for its full height, and a five-story building required a sixteen-inch exterior wall thickness in first floor walls that supported twelve-inch-thick walls above. Thickness increased to twenty inches in the first floor of an eight-story building. Such thickness required another four-inch increase where spans between walls exceeded twenty-six feet. Interior bearing walls were mostly eight inches thick, with lower wall portions at twelve-inch thickness for buildings of more than four stories. Very plainly, the large amount of buildings for residential and smaller commercial purposes required enormous amounts of brick masonry. With the foregoing, it is now appropriate to state the position of the Hudson River brick industry with respect to this prodigious growth in demand, using the year 1890 because of its occurrence at the start of the two decades of New York City's greatest growth. A critical document[22] of that year shows Hudson River brickmakers comprising seventy-six percent of all brick suppliers to the New York market, with the inevitable result of positioning the Hudson River as the brickmaking center of the world by 1900.

Wages

Less variable than brick prices were workers' wages and the length of the workday. The 1880 Products of Industry Census showed that Ulster County brickyard wages varied between $1.50 and $2.00 per day for skilled labor and between $1.00 and $1.40 per day for unskilled labor. The workday varied between ten hours and fourteen hours (one yard only), with the average at twelve hours. Cordts and Hutton paid at the top rate for both skilled ($2.00 per day) and unskilled ($1.40 per day) labor and had a ten-hour workday. Two other Kingston yards also were singular on the river for paying that same top wage rate—A. S. Staples and Nathan Nickerson. However, Staples got his money's worth with a fourteen-hour day, while Nickerson had the average twelve-hour day. The Haverstraw skilled wage varied from $1.75 to $1.85 per day and most yards had a twelve-hour day, with a few at ten hours. Comparison with other industries is constructive. Finished metal products industries (stove manufacturers and foundries) in Troy, still a leading American manufacturing center in 1880, paid skilled worker wages between $2.00 and $3.00 per day, with an average of approximately $2.40 per day. Unskilled wages in the same Troy industries varied between $1.12 per day and $2.00 per day, with an average around $1.50 per day. Those same Troy industries generally had a ten-hour workday. In 1880, the cement and quarry industry of Ulster County paid skilled wages between $1.50 and $1.75 per day and unskilled wages between $1.00 and $1.25 per day, with a ten-to-twelve-hour workday. It should be noted that Henry Ford's wages at the Highland Park plant, in 1914, were considered quite generous, at $5.00 per day for an eight-hour day.[23] A phenomenon of the personnel scene was the large number of Irish foremen, whose indispensable presence deNoyelles testified to: "If I were to choose one ingredient which spelled successful brickmaking in Rockland County, it would have to be the Irish foremen. . . ." The large number of Irish immigrants working in the Downriver yards during the mid-nineteenth century occasioned this, when the large influx of newly immigrated—and very poor—Irish living in the city met the great need for manpower at the Haverstraw yards. Coincidentally, the Hutton Company also had foremen of Irish descent—with a considerable representation—as late as the mid-twentieth century. They were frequently referred to as being among the 'key men.'

Labor Relations

Inevitably, there was conflict between workers and management over the various conditions of work and compensation. Management pressed for a twenty-five thousand brick daily pit production by each machine, with workers requiring a limit at twenty-two thousand five hundred bricks. Downriver, disagreements over this contract could explode into outbursts of sabotage that took the form of toppling hacks of brick and extended to damage of machinery and equipment. Some Haverstraw owners attempted to break strikes by importing French Canadian workers in 1877. The striking workers reacted violently, which led to a sheriff's request for aid from the governor in the form of two National Guard companies. Other strikes were occasioned by workers' representations for pay increases, usually in the range of ten to twenty-five cents per day. Genuinely organized labor representation had limited beginnings. In the 1880s, a national organization, the Knights of Labor, established a presence in the valley at Verplanck Point (Westchester County). In 1886, the Knights successfully bargained for advantageous conditions for wages, hours and production quotas. However, in 1887, brickmakers refused an effort to meet with workers for the purpose of recognition of union representation, and this resulted in a sixteen-month strike that brought on strikebreakers. A request to Governor Hill for militia—a request that was denied—testified to the growing strength of the labor movement. The manufacturers' reaction was typical for the time.[24] Strikes could be limited to relatively isolated communities like Verplanck or extend through entire nearby brickmaking centers, such as Haverstraw, where an 1898 strike produced cooperative action, led by John B. Rose, that included assessing all manufacturers on the river for the purpose of building a sizeable fund for the benefit of struck manufacturers, thereby allowing them to hold out against the strikers. Increased wages were a very real problem presented to the manufacturers, insofar as wages were the largest cost in brickmaking. The relentless pressure on brick prices in the New York market deprived the manufacturers of any flexibility in pricing—and was guaranteed to create conflict over variables such as working conditions and wages. Industry-wide labor organization was delayed until the 1930s. Greatly increased productivity allowed significantly increased wages—but with far fewer workers.

Making a Living in the Off-Season: Ice Harvesting

Brickmaking was a seasonal business, with some yards open only six months and some, like Cordts and Hutton, operating for eight months. The average was around seven months. Employers frequently extended interest-free cash advances to their workers during the shut-down times, on the basis of a handshake. The close of navigation on the river due to ice, the freezing of the clay banks, and inclement conditions for drying necessitated the seasonal character of the business. There existed a pressing need to fill the winter with employment for the brick workers. That need was satisfied to a considerable extent by the manpower requirements of the ice harvesting industry, with New York City again providing a voracious demand for yet another Hudson River product. The ice business may have had its origins as early as 1830 around Haverstraw. Rockland Lake, perched on a height of land immediately above the river, became a principal source of ice from pure water—and required an inclined railway for access to river transportation. Greene County became a principal location for the ice business. It began shipments in 1847 and had thirty-five ice houses by the mid-1880s, ranging in capacity from twelve thousand to sixty thousand tons, with this activity centered near Coxsackie and New Baltimore.

Very quickly, the ice business boomed and the Hudson River became punctuated with enormous wood ice storage buildings—'ice houses'—some longer than a football field, fifty feet high, heavily insulated with sawdust in their double-wall construction and equipped with steam-driven hoists for storing ice inside the buildings. The ice business had its own dominant figures, such as the preeminent Knickerbocker Ice Company. The first ice houses appeared around 1850, while barns and various other kinds of structures were used before that time.

There was an old pejorative expression that was applied to any building that was big and featureless—'looking like an ice house.' They were painted a light color—often yellow—and you couldn't possibly miss 'em. A maximum of two harvests could be expected from the river ice fields, due to tidal disruptions, allowing lakes to provide harvests well in excess of the river locations.

After snow clearance and initial scoring of blocks by horse-drawn implements, the ice was cut directly from the river with seven foot long, manually-operated saws and steered through channels in the ice to the ice houses' endless chain conveyors (powered by 25 to 35 horsepower steam engines) that A. and B. Wiltse of Catskill perfected. A small industry arose in the ice harvesting equipment business, also centered on Greene County, where the blacksmith-turned-manufacturer Herman Dernell of Athens became a leader in the field, exporting ice harvesting equipment to England, Sweden, Norway, and Russia. Bear in mind that the ice didn't necessarily wind up as ice cubes in the drinks, as it was principally used in ice boxes for cooling foods.

The ice business was a good fit with the brick industry and many brickmakers were also in the ice business. Immediately north of the Terry yard, the Dwyers' Kingston Brick and Ice Company and the Hanrahan Brick and Ice Company (at Ulster Landing—immediately North of East Kingston) made the connection between brick and ice eminently clear. Powell and Minnock were in the ice business for ten years, at Coeymans, before they made bricks. A panoramic series of photographs[25] of the river's shores in 1906 shows ice houses sometimes resembling an almost continuous cliff, as at East Kingston.

Brick Transportation

The change in the form of transportation of bricks to the metropolitan market is a story in itself. Bricks were shipped on the decks and in the holds of sail-driven sloops and schooners, through the 1880s, with a schooner carrying approximately eighty thousand. A limited number of sail-powered vessels continued to carry brick cargoes until 1910. With the advent of steam-powered tugboats, brick transportation changed dramatically, with the Cornell Steamboat Company of Rondout virtually monopolizing towing services. The manufacturers, chafing under limited shipping options, briefly contemplated the possibility of forming a competitive towing company.

The brickmaking Washburn Brothers of Glasco mounted the only real challenge to Cornell dominance with a fleet of seven tugs that Cornell bought out in 1884. This purchase included two brand new iron-hulled tugs that had no equal in the Cornell fleet of that time. The greater speed of the newly economical steam-powered tows also removed the handicap of distance from the market that was a severe disadvantage to the Upriver manufacturers and a distinct advantage to the Downriver brickmakers.

A similar change in transportation advantage, in the 1960s, provided a poignantly ironic and fatal condition for the Hudson River brick industry as a whole. Large capacity scows, or barges, carried three hundred to three hundred and fifty thousand bricks, with barges assembled into groups of twenty to thirty—termed 'tows.' The barge captain had considerable responsibility in the movement of the brick cargoes to the market. Daniel deNoyelles depicts barge life, with the live-aboard captain and family, meticulous shipshape care of the vessel, and awnings for shading the barge crew while working. On-board crews of six men, for unloading the cargo at the city dock, slept in a forecastle. Union-organized stevedores forcibly ended the days of live-aboard crews in 1910. This initial New York City waterfront association of stevedore union organization with dockside brick-handling eventually extended, industry wide, to all brickhandlers at the yards as well.

Barges frequently left the yard with no knowledge of the exact market price that was to be obtained upon arriving at the city berth. This gave the captain a pivotal position as both salesman and business representative, and they were renowned for their prowess at speedy mental computations. Captains were the elite of all the personnel and frequently were co-owners of barges with the brick manufacturers, sharing freighting receipts. The higher salinity of the lower river gave the Downriver manufacturers another advantage over the fresh water Upriver yards—a large reduction in winter icing. This allowed the Downriver yards to take advantage of both very late and early construction season brick needs that could create substantial short term price increases.

Rail Shipments

Rockland County yards made limited use of rail shipments, apparently shipping to New Jersey markets. At least one yard at Thiells (immediately west of Haverstraw) that did not have direct access to water transportation used rail transport. Other companies included the Peck Brick Company of Haverstraw, with its own siding, that connected to the West Shore and Erie Railroads, as well as a Cornwall yard (Orange County) that shipped via the New York, Ontario and Western Railway. The totally landlocked Lowe yard, at New Paltz, also shipped solely by rail, via the Wallkill Valley Railroad that connected to the Erie Railroad at Goshen. Columbia County's Cary yard at Newton Hook was unique in its ability to access the late-nineteenth-century Boston market by rail. It had its own rail spur from the main line New York Central tracks which separated its clay banks from its other manufacturing processes on the river.

Finite Clay Resources

The clay resources were not inexhaustible. The first major evidence of their finite nature appeared in Haverstraw during the 1880s with the commencement of clay excavation below the river's water level. In 1883, Daniel deNoyelles undertook a sizable clay excavation that was both directly adjacent to the river and below the water level at his South Haverstraw yard with its three quarters of a mile of shoreline. This could only be accomplished by constructing a continuous barrier to the entry of river water. Such a structure is called a cofferdam. The cofferdam formed the east (adjacent to the river) side of the excavation area, on its long north-south dimension of two thousand feet. Land bordered the west side and part of the north and south sides of the excavated portion that was approximately two hundred fifty feet in its east-west dimension—it was a rectangular area (total size of two thousand by five hundred feet or eleven and one-half acres) partially cut into the shore.

Figure 33. Bird's-Eye View of Haverstraw and the Stony Point District by L. R. Burleigh, 1884. Inset: deNoyelles property and cofferdam

References from Haverstraw map:
A cofferdam
K clay pit
O narrow-gauge railway
P Minisceongo Creek
(Note: J. R. Burleigh, trained as a civil engineer, was known for the accuracy and detail of his depictions)

References for bird's-eye view:
bm brick machine
by brickyard(s)
cy covered drying yard
dy open drying yard
ex excavation (clay pit)
fb foot of original bluff
ks kiln shed
py palette yard
R ridge created by excavation

A contemporary illustration (Figure 33) shows the cofferdam projecting east (into the river) from the line of the main dock for a distance of approximately one half of the cofferdam's east-west dimension with the other half being cut into the shore land. The illustration shows the abrupt rise of the High Tor mountainside immediately to the West and the railroad track very nearby and up the slope. The cofferdam structure, fully exposed to ice movement on the river, lasted for twenty-five years and required constant repairs as well as seepage removal by two pumps at fifteen hundred and eighteen hundred gallons per minute. The cofferdam excavation ceased to be used after the pumps could no longer remove the water at a rate needed to keep the excavation dry because of the depth below the river level.

Two Haverstraw yards—Fowler and Excelsior (both under Fowler ownership)—used big, floating continuous-chain bucket-dredges to load rail-mounted dump cars that were floated out to the dredges on barges, obtaining clay that was below the surface of the river itself (Figure 36). With a water depth of only four feet, a twelve-hundred-foot-long access channel had to be dug through the shallow flats. At the Fowler yard, the loaded rail cars were then rolled off onto trestles, one of these having been five hundred feet long and thirty feet high, requiring steam-powered cable traction for the movement of the cars up to that height. The height of the trestle was required to enable large amounts of clay to be dumped onto stockpiles before using it, in order to drain it of the river water with which it was saturated—a process that required at least forty days.

An initial reason for dredging was to upgrade the quality of Excelsior's mix, since the river clay was of considerably higher quality than the dry land material. This landside source was eventually given up. Both yards finally drew all of their clay from the river by dredging. The cofferdam and the dredging represented ingenious and resourceful ways to deal with the very serious problem presented by diminishing clay resources—but they were also expensive, in relation to yards that did not have to resort to such methods. In 1940, the fate of the Excelsior yard and the Hutton Company became intertwined with some finality. In addition, allotments of Haverstraw clay lands were so dense and excavations so tight against property lines that legal challenges to encroachments were frequent. Surveyors were kept busy checking the property lines and there were frequent daily inspections of those limits. Any clay from an adjoining property that collapsed onto adjacent land was an accidental gift to the lucky recipient.

A Community in the Grip of an Industry

Between the village of Haverstraw and the river lay the Bowline clay lands, named after the point on the river's shoreline. A wide strip of shoreline—today appearing like an isthmus—was kept intact, as the location of the brick machines, drying yards, and kiln sheds. That strip of land also acted as a natural barrier to the river. Several brickmakers subjected the area between this strip and the village to intensive excavation, consequently lowering the land surface below the

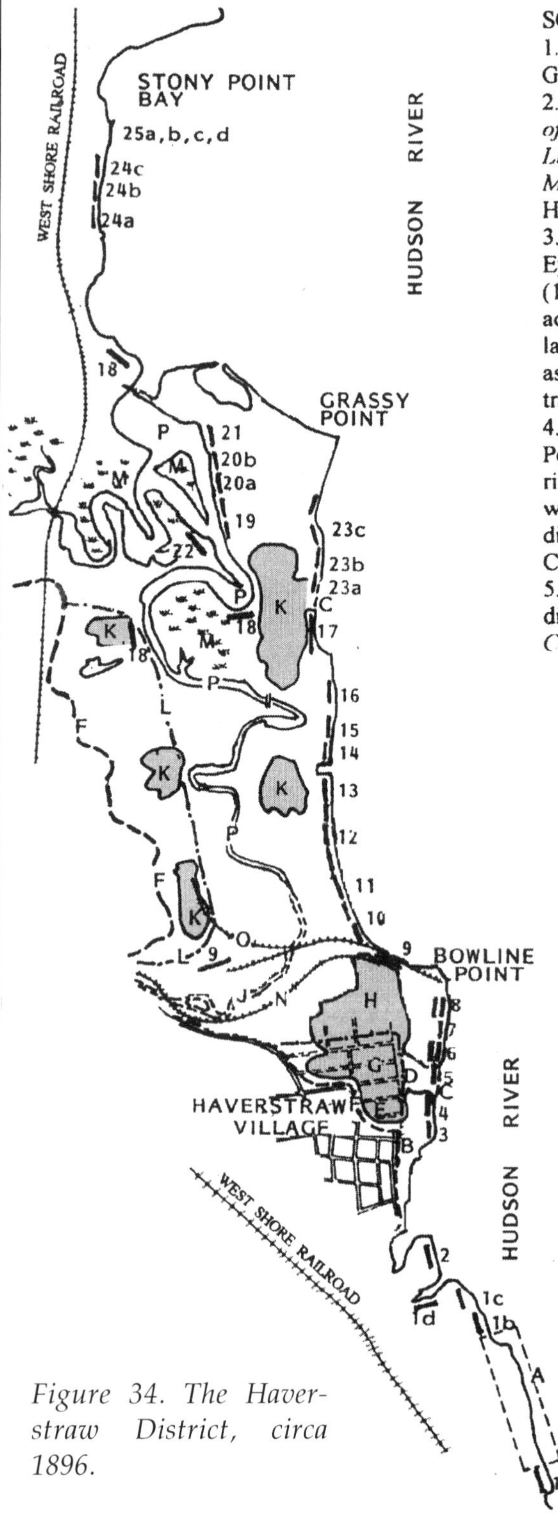

SOURCES FOR MAP INFORMATION

1. Map shows existing topography on U.S. Geological Survey map of 1964
2. Brickyard Locations: from *Insurance Maps of Haverstraw* by Sanborn Map Co. (1896), *List of Manufacturers Tributary to New York Market* (1890), and *Clay Industries of NY* by Heinrich Ries (1890).
3. For contemporary landscape features: Birds Eye View of Haverstraw by L. R. Burleigh (1884); such depictions are reputed for accuracy as a record of contemporary landscape, using atlases, maps and land records, as well as on-site observations. Burleigh was trained as a civil engineer.
4. Bluff crest between Haverstraw and Grassy Point: in mid-1890s, one quarter mile from river, per H. Ries in *Clays of New York* (1900), with this stated distance appearing as a discrepancy with relation to Minisceongo Creek (bluff was west of Creek).
5. "Most of the yards along the river are digging their clay below tide level" (H. Ries in *Clays of New York* – 1900)

TOPOGRAPHY FEATURES

- A DeNoyelles Cofferdam — — —
- B Existing Crest of Bluff (assumed to be in original location; shown — — —)
- C Original Shoreline (shown— — —)
- D Original Crest Of Bluff (as extension of existing original bluff crest to South - shown —- · —)
- E Site of collapse of 1906
- F Existing Crest Of Bluff (shown — —)
- G Proposed Street Plan (from F. A. Davis Atlas of 1876; shown = = =)
- H Bowline Pond (clay pits shown shaded)
- J Minisceongo Creek in 19th century (shown ≈≈≈)
- K Clay Pits (shaded)
- L Original Crest Of Bluff (from Birds-Eye View of 1884; shown — - · —)
- M Wetlands
- N Siding Tracks For NY & NJ Railroad
- O Narrow Gauge Railroad Tracks of Peck Co.
- P Minisceongo Creek

0 ½ 1
miles

Figure 34. The Haverstraw District, circa 1896.

MAP NOTES
HAVERSTRAW BRICKMAKING DISTRICT
Owners and Operators (from Sanborn Insurance Map of 1896; dates and names from other sources as indicated)

1. deNoyelles Estate; a. Henry Christie, operator (Malley, successor operator); b. deNoyelles, operator; c. Shankey, operator; d. Patrick Lynch, operator
2. Denton Fowler, owner, operator
3. Excelsior Brick Co., owner, operator
4. Theodore Frederick, owner; Donnely & Son, operators
5. J. E. Eckerson, owner; Gormley & Co., G. Sherwood, Wren & Co., operators
6. Archer Estate, owner
7. F. Eckerson, owner; Philip Goldrick, operator
8. Mrs. Snediker, owner; Snediker Bros., operators
9. Gordon & Theodore Peck, owners, operators
10. Archer Estate, owners; Archer Bros., operators
11. J. B. Allison, owners, operators
12. Rowan & Scott, owners, operators
13. J. B. Allison, owner; Wood & Allison, operators
14. Mrs. Reid, owner; Cahill & Carroll, operators
15. Washburn & Fowler, owners, operators
16. Lilburn Estate, owner; Sliney & Hettlinger, operators
17. Warrall & Palmer, owners; Washburn Co., operators
18. Unnamed brickyard (shown on bird's-eye view of 1884, L. R. Burleigh, publisher)
19. Palmer & Co., owner; James Morrissey, operator
20. Felter Bros., owners; a. Farley & Barnes, operators; b. Patrick Brophy, operator
21. Lilburn Estate, owner Thomas Malley, operator
22. Philip Goldrick (from 1887 Sanborn Insurance Map)
23. Grassy Point operators named in *Mfrs Tributary to NY Market* of 1890 (exact locations uncertain): a. T. C. McGuire; b. Kelley & Byrnes; c. Richard Crowley
24. Lilburn Estate, owner; a. McGowen & McGovern, operators; b. Freeman & McGuire, operators; c. Hurley & Brophy, operators
25. Stony Point operators (named in Ries, *Clay Industries of NY*, 1895, exact locations uncertain): a. Allison, Wood & Allison; b. Reilly & Rose; c. Reilly & Clarke; d. C. A. Marks

river's water level. The Donnelly yard's excavations extended forty-five feet below mean tide level by 1900. One half of the eventually excavated area was shown on an atlas of 1876 as a proposed extension of the village's street and residential lot pattern, an area of approximately thirty acres, with the surface of the ground varying from twenty to forty feet above the lower river-level ground surface.

In its final form, the excavation went from the crest of the original bluff, eastward toward the village, for a distance of approximately seven hundred to fourteen hundred feet, the distance varying with the top of the excavated bluff that skewed to the west.[26] This is confirmed by the 'bird's-eye view' of 1884 (Figure 33)and insurance maps that show a precariously narrow, fingerlike promontory of unexcavated land, with its top surface at the village's land level, projecting east from the village toward the river, and with precipitously steep banks that defined this promontory. Those banks plunged into the clay pits on each side of that promontory, with houses of die-hard residents arranged along the top of this precarious man-made ridge that eventually disappeared.

Additional evidence is provided by a U.S. Geological Survey map that indicates the original bluff line. This map is used as the basis for the map (Figure 34) that shows the extent of the Haverstraw brickmaking district. There were be dire consequences from those excavations in the early twentieth century.

The balance of the Bowline clay lands, originally close to the river level,

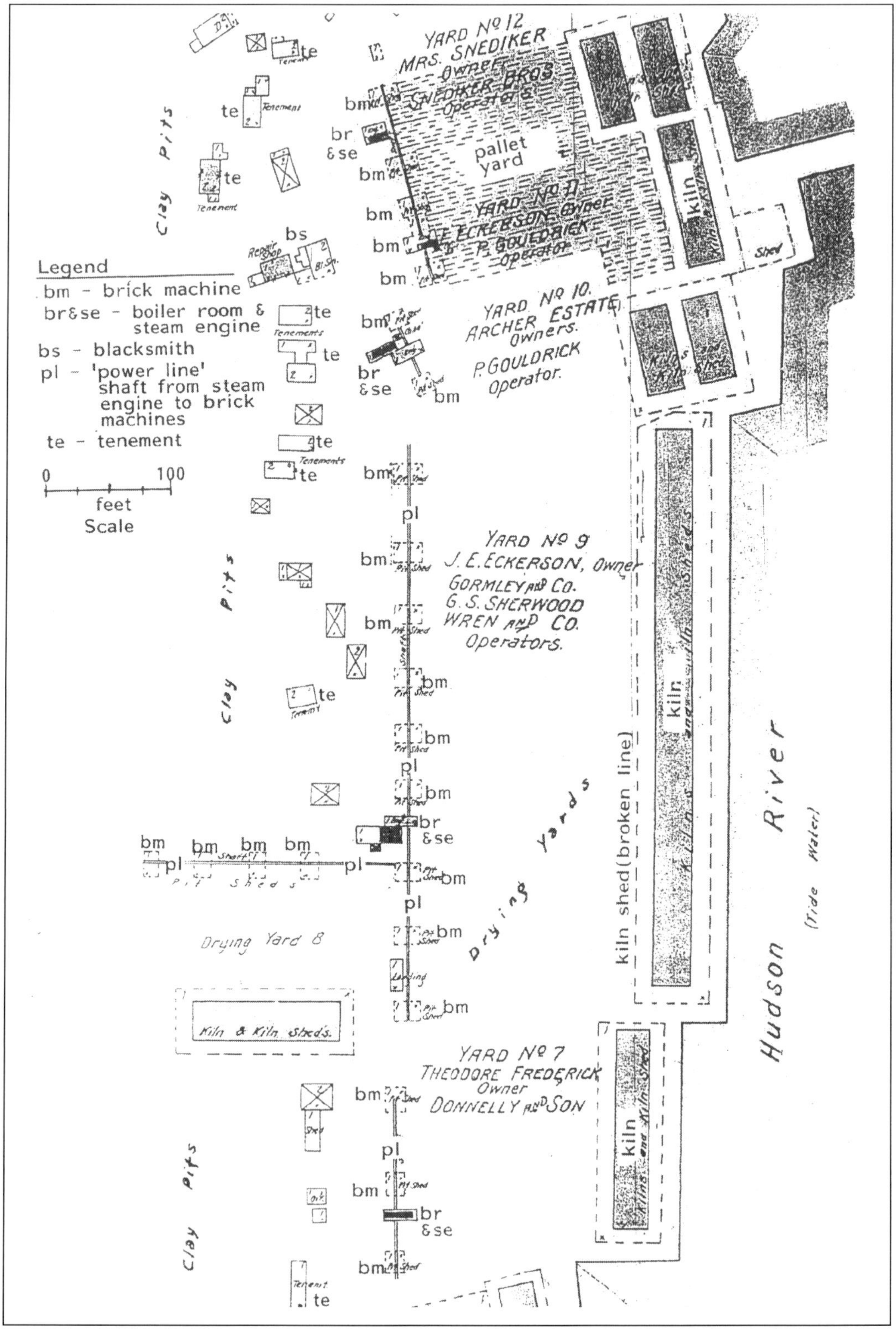
YARD № 12
MRS. SNEDIKER
Owner
SNEDIKER BROS
Operators
pallet yard
YARD № 11
J. E. ECKERSON Owner
P. GOULDRICK
Operator
YARD № 10.
ARCHER ESTATE
Owners.
P. GOULDRICK
Operator.
YARD № 9
J. E. ECKERSON, Owner
GORMLEY AND CO.
G. S. SHERWOOD
WREN AND CO.
Operators.
YARD № 7
THEODORE FREDERICK
Owner
DONNELLY AND SON
Legend
bm - brick machine
br&se - boiler room & steam engine
bs - blacksmith
pl - 'power line' shaft from steam engine to brick machines
te - tenement
0 100
feet
Scale
Clay Pits
Drying Yard 8
Drying Yards
Kiln & Kiln Sheds.
kiln shed (broken line)
kiln
Hudson River
(Tide Water)
Pit Sheds
Shaft
Repair Shop
Tenements
Shed

were subjected to the same intensive excavation to levels well below the river's surface. Significant seepage into the clay pits, all below the mean tide level at this location, also required extensive pumping for water removal.[27] 'Corduroy' road surfacing (made from tree trunks placed side by side) for access to the pit bottoms was also required. This was subject to rapid rot, laborious movement of clay carts up the inclined roads, and increased expense. Later in the twentieth century, the narrow strip of land along the river was breached with dynamite to allow permanent flooding of the Bowline clay lands with river water and the formation of what is now called Bowline Pond. That nearly fifty acres of quiet water stands today as mute testimony to the intense activity and wholesale change of the landscape that once took place there. Keep in mind that men using hand shovels removed all of this enormous quantity of material.

Immediately to the north, the lands between Haverstraw and Grassy Point were subjected to the same intensive excavation for more than one mile. The same bird's-eye view of 1884 (Figure 33) shows extensive excavations, these varying from twenty to forty feet in depth below the mean tide level of the river, leaving evidence today in the form of large ponds. Ries[28] stated that there were no clay deposits above tide level between Haverstraw and Grassy Point, and Fowler and

Figure 35, facing page. Four Haverstraw Brickyards. From the Sanborn Insurance map, 1896, showing density of brickyards, clay pits, and tenements.

Figure 36, above. Dredging Clay From The River at Haverstraw. Clay cars are mounted on rails on deck of barge, with tug at left towing clay car barge. (Courtesy of New York State Archives)

Figure 37. The Charles A. Shultz Brickyard at East Kingston. Copper plate etching circa 1885; the slip in which the two schooners (the Hannah Brown *and the* Anne E. Webb*) are docked exists today; both of those vessels carried bricks to Providence, R.I. (Courtesy of Herbert L. Shultz)*

Washburn had to go some distance (more than one quarter of a mile) west to obtain clay from a fifty foot high terrace; other yards obtained clay from the same terrace. Ries also stated, "Most of the yards along the river are digging their clay below tide level." The biggest of those clay pits, now a marina, was used for the construction of large foundation caissons for building the Tappan Zee Bridge foundations (after opening a channel to the river). These were floated down to the bridge site. Nowhere else on the Hudson River did there exist such a compressed activity of brickmaking as at this portion of the river's shoreline. It characterized the Haverstraw scene of an almost continuous sprawl of clay excavations, drying yards, and kiln sheds, running for over three and one half miles from South Haverstraw, along the Haverstraw shoreline, and up to Stony Point (Figures 33 and 34). That was the look of the brickmaking center of the world in the late-nineteenth century.

Upriver: New Developments and New Arrivals

Nature could be fickle in various ways, as with the Columbia County Brousseau yard at Schodack (opened in 1870). The silting of its access channel to the river forced this yard out of business.

Notable additions to Upriver manufacturing during the post-Civil War period included Charles A. Shultz who established a yard at East Kingston in 1876 (Figures 37 and 38). His father had emigrated from Bohemia in the 1830s and had various businesses in the United States, including a brickyard in New Jersey. Charles Shultz would hold both U.S. and Canadian patents for a mold design that produced a decorative finish on the bricks. The East Kingston yard also became a principal manufacturer on the river. The plant left Shultz's ownership in 1916 and continued in operation, bearing the Shultz name, until World War II.

Alva Staples's forebears included a member of the Irish Parliament who emigrated to the New World in 1745. Born in Marlboro (Ulster County) in 1832, Alva began his career as a storekeeper, followed by entry into the brick business north of the (Kingston) Steep Rocks Terry yard (Figures 39 and 40). His son, Alva Jr.

Figure 38, top left. Boiler Flue. Charles Shultz brickyard(as shown in Figure 37, left end of picture); it is in immaculate condition after almost 120 years; such structures are subjected to extremes of weathering, due to their free-standing position. (Photo by author, 2000)

Figure 39, top right. Terry Brothers Brickyard, Kingston, at Steep Rocks, circa 1880. An empty barge is being berthed by a tug; a large supply of wood kiln fuel is stacked beside the road. (Photo in collection of Hudson River Maritime Museum; furnished by John F. Matthews)

Figure 40, middle. Terry Brothers Brickyard, Detail of Figure 39. Note 'mold setters' who are 'dumping' bricks from molds in drying yard.

Figure 41, above. The Brigham Brickyard at East Kingston. This was a unique bi-level yard with kiln sheds on upper and lower levels and an inclined railway (just visible at right end of lower shed) for transporting finished and dried bricks to lower level. Brigham's extensive clay banks were at the upper level. In 1913, this yard had 127 workers and was one of the biggest on the river. (From collection of John F. Matthews)

married Cora Terry, of the Steep Rocks brickmaking Terry family, subsequently starting the Malden yard (immediately north of Saugerties) in 1905, which continued operation until 1958. The proximity of the Terry and Staples yards allowed this "brick marriage" to be easily arranged. The two firms did, however, retain their separate ownership, with the eventual closing of the Staples's Kingston yard after opening the Malden plant. Like William Hutton and John H. Cordts, Alva Staples did not limit his business activities to brickmaking, but also engaged in extensive real estate transactions.

The entry of George W. Washburn and his brother John Tyler into brickmaking at Glasco (south of Saugerties), as Washburn Brothers in 1867, has already been noted. The two Washburn brothers had an older brother, Uriah, who made an 1855 entry into the brick business in Haverstraw, at age twenty-six, and that business became one of the largest Downriver manufacturers. Attracted by his brothers' Upriver success, Uriah also established a yard, located in the town of Ulster, immediately south of the present Kingston-Rhinecliff Bridge (now the town of Ulster Park). That yard's brand—UF&JTW—indicates that his younger brother John Tyler (a partner in the Washburn Brothers yards at Glasco) was also a partner in this business.

Elisha Brigham moved to Kingston from Connecticut in 1849, initially in the storekeeping business and subsequently starting the Kingston and Rosendale Cement Company, followed by the Hudson River Cement Company at East Kingston. His sons, Henry and William, established a short-lived brickyard at Creek Locks (the Rondout Creek terminus of the Delaware and Hudson Canal), followed by an 1891 move to establish the East Kingston yard. That yard eventually had an annual production of forty million bricks, making it one of the biggest on the river (Figure 41). It employed two hundred twenty-five men and continued operations into the 1950s. In 1906, it was also the second yard on the river to end the ancient practice of drying brick by using natural atmospheric conditions, and instead used an early type of steam-heated drying.

Other Ulster County Brickmakers

One of the two greatest concentrations of Upriver brick manufacturing lay in Ulster County, between Port Ewen and Malden (approximately fourteen miles), with thirty-two yards listed in C. E. Hall's inventory of 1905. Of these, twenty-two can be accounted for today by visible remains—significant amounts of brick rubble on the ground. Between East Kingston and Glasco, there were more than forty yards within a distance of less than seven miles over a period of seventy years. Of all of this manufacturing activity, there are virtually no remains except for the brick rubble strewn on top of the ground and along the river shoreline—and even such evidences do not reveal all of the yards' locations. Inventories (See Appendix–Lists of Manufacturers) can be used in order to obtain brands and complete listings. The Hutton Company and Powell and Minnock stand (in 2002) as the only extensive remains of plants from the past on the entire river, along with graceful nineteenth-century brick boiler flues at the sites of the Terry and Shultz yards at East Kingston and the limited remains of the Staples plant at Malden. Because most of the plants' principal structures were constructed of wood, all of those vanished long ago, and the jungle has closed in. It is worth noting the generally favorable comments on creditworthiness and honorable business behavior of the Ulster County brickmakers, as reported by R. G. Dun and Company (predecessor of Dun and Bradstreet) during the 1870s and 1880s. An 1876 entry regarding Cordts and Hutton, in the shorthand of the credit-rating trade, stated: "Good firm, enjoy gd. crd; easy in money matters"; and in 1879: "Do very good bus. Good strong firm here & stand well in every way."

Southern Dutchess County Brickmakers

The other great concentration of Upriver manufacturers was in the Beacon district of Dutchess County, which extended from Chelsea (three miles north of Beacon) to Dutchess Junction immediately south of Beacon. Along that six miles of shoreline, a total of thirty-eight yards existed at the end of the nineteenth century. Brick manufacturing originated at Fishkill Landing (Matteawan and Fishkill Landing were incorporated into the city of Beacon in 1913) during the 1830s—with John Gowdy and Isaac Brinckerhoff (of Dutchess Junction) using pit wheels for tempering and hand-operated presses for molding. In 1847, the tracks for the Hudson River Railroad (renamed New York Central Railroad) displaced some of the earliest yards. In 1852, Tomas Aldrige acquired the Brinckerhoff's Dutchess Junction property and eventually became a major producer (daily capacity of four hundred eighty thousand bricks).

In the early 1880s, there were seven yards at Fishkill Landing alone. During this time, two of the largest yards on the river emerged near Fishkill Landing, the Dennings Point Brick Works and the Brockway Brick Company. Dennings Point, a south-dangling peninsula (immediately south of Fishkill Landing) known earlier as 'Presqu'Ile,' was a favorite stopping place for George Washington while visiting a unit of the Continental Army camped nearby in the early 1780s, tying his horse to one of two great oak trees that stood for a century after that event until a storm blew them down. Owned at the time of the Revolution by the Verplanck family, the son of a member of Washington's staff (Adjutant-General William Denning) purchased the property. Late-nineteenth century ownership of the point fell to Homer Ramsdell, the prominent Newburgh promoter and financier of regional railroads. Ramsdell had accepted a mortgage on the point and some adjacent lands issued by the Boston, Hartford and Erie Railroad, as security for the railroad company's purchase of Ramsdell's Newburgh-Beacon ferry and other Newburgh properties. The railroad company defaulted on their payments, Ramsdell took over the point and then went into the brick business in 1881, building it into one of the major producers on the river.

Thomas M. Wiley made Ramsdell's brick machines. Wiley was a Cornwall manufacturer who learned the brick machinery business while working with

Nathaniel Adams on the development of his pioneering machines during the 1830s. A sixty-horsepower steam engine, built by William Wright of Newburgh, powered the machinery. Initially, this yard's clay was taken from the point itself, subsequently from land on the east side of the railroad tracks, and later supplemented by clay that was dredged from the river. Sufficiency of clay at this yard became an ongoing problem over the years and came to total exhaustion of clay resources in the late 1930s. A notable practice at this yard was in limiting the height of kilns to fifty-four courses, "excessive height in piling a kiln being a grievance which often causes men to quit work at inconvenient times."[29] The red shale on this property was finely ground for use as a coloring agent. A pre-existing paint company had used the same material as a pigment. The Brockway yard, one and one half miles north of Beacon, was started with Edwin Brockway's 1886 purchase of seventy acres of land that had been the 'country seat' of William Mortimer. This yard became one of the two or three largest producers on the river. This company had a community named after it (Brockway), in the same manner as naming Roseton after the Rose Brick Company.

Edwin Brockway was born in Glasco (south of Saugerties) and moved to Haverstraw with his family in1846. He was a brick manufacturer in Haverstraw until his acquisition of the William Mortimer estate, where he intended to retire. Acceding to his sons' desire to enter the brick business, he also acquired a New York State grant to seventy-five acres of river surface west of the shoreline in 1886, including rights to the underlying land. With the shore line too close to the railroad tracks to allow brickyard operations, sufficient fill was placed on the newly acquired submerged land in 1887 to allow commencement of brickmaking operations in 1888 on the filled land. Initial kiln masses caused several successive settlements that required additional fill in order to level the surface under the kiln shed. Further filling operations in 1905 and 1906, using "dry garbage" from New York City, provided additional westward extension of the land into the river. The latter fill was the cause of settlements of the post-World War II tunnel kiln that caused kiln car derailments.

In 1898, the Brockway Brothers Company also commenced brickmaking at Dutchess Junction. The Brockway yard owned sixteen brick machines by 1910. It employed four hundred and had a capacity of one hundred million bricks per year, thereby making it a contender for largest producer in the Hudson River brick industry. Having insufficient on-site sand resources, in 1897 the Brockways acquired one hundred and fifteen acres of sand property at Plum Point, New Windsor (once known as the Verplanck Sand Property), on the west shore of the river. This property is now the Kowawese Unique Area park, bearing an historic marker stating that this property was also used for sand excavation by the Dennings Point brickyard.[30]

Adjacent to the Brockway property, the death of William Verplanck brought on leases on the lands of his estate that included names of consequential manufacturers like William Lahey (owner of several yards), Clayton Bourne, and

Figure 42. 'Edgewood Terrace' —the Ponckhockie (Kingston). Residence built by John H. Cordts in 1870 on a height of land southeast of the Cordts and Hutton brickyard; the lady with the parasol is Mrs. John N. Cordts; seated, in black, is Elizabeth Lowe Cordts (wife of John H. Cordts); the infant is Florence Cordts, daughter of Mrs. John N. Cordts. (Courtesy of Matilda Cordts Davenport and the Cordts family)

William Aldridge (in addition to his yard at Dutchess Junction), as well as Thomas Dinan (also operating at Grassy Point) and O'Brien and Vaughey—thereby making Brockway a very active center of brickmaking.

Another notable occurrence in the region's brickmaking history was when the two daughters of the deceased owner inherited the William Budd yard at Fishkill Landing. 'The Misses Budd' not only increased the plant's output but reputedly made the first use of electric power for the operation of Hudson River brick machines.

Roseton

Another manufacturer who commenced operations of one of the most progressive and most long-lived operations on the river was Juan Jacinto Jova, a Cuban sugar importer and broker in New York City who represented the family's plantation in Cuba as well as other Cuban sugar interests since just before the Civil War. Juan Jova began his business career in the city as a young man, in 1874 buying a large Greek revival granite house (Danskammer) on a sizeable piece of land four miles north of Newburgh. After leaving the sugar business, it was there that he began brickmaking in 1884, at what was to become known as Roseton. Readily enough, the locals took up the aspirated first syllable Spanish pronunciation of the name (HO-va).

The Jovas had their origins in Catalonia, the northeastern province of Spain, which was known for its industrious and resourceful people, as well as being a principal source of entrepreneurs for the Spanish colonies. The homestead remained up the hill at Danskammer—similar to the proximity of the John H. Cordts mansion to the Cordts and Hutton yard.

Another Cuban family of Catalan extraction, the Abreus, would became associated with the Jovas and their brick business through marriage. The Jova brand—J J J (the Triple-J)—had the most memorable graphics of all, with the exceptions of Haverstraw's Diamond brand with its stark parallelogram shape, and Barnes's XXX brand. The Jovas and their plant were principal players in the history of the industry.

Coeymans

Toward the north end of the Upriver region, at Coeymans (ten miles south of Albany, on the west shore), two principal yards emerged at the end of the nineteenth century. Conrad Suderley (born in Saugerties in 1855) started Sutton and Suderley. Suderly learned the brick business at the Washburn Brothers' Glasco yards, then married the daughter of John Sutton, brickmaker. Their son, John, married a daughter of the Bronck clan, thereby acquiring the brickyard property. The Broncks held seventeenth-century letters patent to sizable tracts of land around Coxsackie (twenty miles south of Albany on the west shore). That Bronck family, beginning with Pieter Bronck (purchaser of the lands from Indian sachems and confirmed by royal patent) who emigrated to the New World from Scandinavia around 1640, was related to the Bronck lineage whose name became immortalized in the naming of the New York City borough of the Bronx. Under Sutton and Suderly ownership was the adjacent Roah Hook Brick Company, operated as a separate business. Powell and Minnock bought out Sutton and Suderly in the late 1950s. In the late-nineteenth century, the Powell family also married into the Bronck clan.

Dr. A. T. Powell, in partnership with Thomas Minnock, went into the ice business in 1896. On the Powell lands were two lessees making brick, Ziegler and Adamo, these men being bought out by Powell and Minnock, in 1906. Thus began operations of the sole remaining Hudson River brick manufacturer that existed into the Third Millennium.

Nothing Stays the Same

An additional change should be noted, by way of the cessation of the Cordts and Hutton partnership in 1887. That obviously successful partnership lasted for twenty-two years and, with John H. Cordts in his late sixties, that gentleman was entitled to retirement. He died in 1891. From 1890 until 1965, the business was known as The Hutton Company. William Hutton became sole owner and president, assisted by his son George. Fifteen years prior to that time, William and his family had moved into what is still known as 'the General Samson House' on 'the Weinberg', now known as West Chestnut Street. The precise origin of the sobriquet for the street name is not clear, although it is literally translated from German as "wine hill" (Kingston had a significant nineteenth-century German immigrant population living just down the hill in Rondout where members of that community operated two prosperous breweries, thus indicating their true choice of beverage). The house is high style Italianate, built of local limestone in 1858, and today is in splendid condition. George, a merit student at Kingston Academy, was listed in an 1883 edition of the Kingston Directory, at age twenty, as a bookkeeper—a skill that qualified him as the company's treasurer. He continued as the company's treasurer and secretary until his death in 1920. William Hutton retained his business address at his Rondout lumberyard, over a mile from the brickyard.

Besides the brick plant, there were other developments on the old Cantine Patent lands that were sold to Cordts and Hutton in 1865. In 1873, to the south of the brickyard and at an upper elevation, John H. Cordts built a splendid brick mansion—'Edgewood Terrace'[31]—that commands one of the most breathtaking views on the entire river. Also to the south of the brickyard, there began a neighborhood of neat frame houses, with a street on the hillside to be named Cordts Street (entered in municipal records in 1870). The building for the John N. Cordts Hose Company (since 1894), a volunteer fire company that is still in existence, was constructed at the top of the first steep hill ascending Delaware Avenue.

This chapter closes with the death of William Hutton in 1897, marking the end of an era for his brick plant and, in important ways, the beginning of the end for the industry as it existed at that time. His ambition to accrue a significant financial competence was essentially realized—an accomplishment that would have been difficult if he had been limited to the brick business. The family monument in Montrepose Cemetery (Figure 43) is a marble sculptural allegory showing three female figures (old age, youth, and maturity) spinning the thread of life. The central figure of youth is a prime specimen of topless young womanhood—good, life-affirming stuff for a cemetery. Without the need for either himself or John H. Cordts to squeeze every dollar out of a hard business and their employees, that circumstance did allow them to pay their men a very decent wage and to have a reasonable workday, by the standards of the time—and also enabled the substantial investment in the plant that gave them an early position of leadership.

Figure 43. Hutton Monument in Montrepose Cemetery, Kingston. Final resting place of William Hutton, who started out as a poor boy in the 1850s; the figures are an allegorical representation of Old Age, Youth, and Maturity spinning the thread of life. (Photo by author 2002)

Notes

1 The percentages for the values of production have not been adjusted for the large inflation rate of sixty percent for the years 1860-1865, per Cost of Living Index of the Federal Reserve Bank of New York. Adjusted for inflation, the Upriver production value increase would be three hundred percent and Rockland County fifty-three percent, in 1860 dollars.

2 Populations:
Greater London, 1880: 4.76 million; New York City: 1.91 million.
Greater London, 1920: 7.44 million: New York City: 5.62 million.

3 Weatherby. *Nearly 300 Years of Moise Quantin and Descendants in America.*

4 Marius Schoonmaker. *The History of Kingston.*

5 Figure 8. Horsepowered brick machine.

6 Uhland. 1879.

7 *Brick and Clay Record.* Feb. 8, 1908.

8 Figures 13 and 14. Hutton Company, circa 1900. Photographs.

9 Watson and Company. Sect. 22 and 23.

10 See Figures 13 and 14.

11 Amberg.

12 Figures 23 and 24.

13 *Brick.* Jan. 1904.

14 *History of Dutchess County.* Frank Hasbrouck.

15 Searle.

16 See APPENDIX section entitled KILN COMPARISONS.

17 From Green's *History of Rockland County.*

18 deNoyelles.

19 This increase must be adjusted for the sixty percent increase in cost of living (per Federal Reserve Bank of New York Cost of Living Index) between 1861 and 1865. A deflated increase would have been closer to 360 percent.

20 *Manufactures of the United States.*

21 Redfield. *Travelers Guide.* 1872.

22 *Brick Manufacturers Tributary to the New York Market.* 1890.

23 *The House of Morgan.* Ron Chernow.

24 "We cannot afford to have our business controlled by dictatorial unions. . . . We appreciate the right of any class of men to band themselves together for mutual protection and advancement, but deny their right to interfere with and disrupt our business because we do not employ a certain kind of labor. . . ." (*Peekskill Evening Star.* Aug. 5, 1967).

25 Wallace Bruce. *Panorama of the Hudson.*

26 This dimension is derived from U.S.G.S. topographic map (Haverstraw, New York) evidence showing location of original face of bluff immediately to the south of Bowline Pond; verified by the proposed street plan shown on the *Combination Atlas Map of Rockland County of 1876.*

27 deNoyelles.

28 Ries, Heinrich. *Clays of New York.* 1900.

29 From *The Story of Brick.* C. E.Hall 1905

30 Transcript of taped interview of H. Mortimer Brockway; letter from Brockway of 23 March 1981 (both in the archives of the Beacon Historical Society).

31 Figure 42. Photograph of Edgewood.

Principal Sources for Chapter 2

Adams, Arthur G. *The Hudson.*
Chandler, Maurice. *Ceramics in the Modern World.*
Cole, Rev. David. *History of Rockland County.*
Condit, Carl W. *American Building.*
deNoyelles, Daniel. *Within These Gates.*
Dobson, Edward. *A Rudimentary Treatise on the Manufacture of Bricks and Tiles.*
Engineering News Record. *Construction Costs.*
Ford, James. *Slums and Housing.*
Green Frank B. *History of Rockland County.*
Hall, C. E. *The Story of Brick.*
Hasbrouck, Frank. *History of Dutchess County.*
Horne, Field. *Greene County: A History.*
Lawrence, W. G. and R. R. West. *Ceramic Science for the Potter.*
McMahon, Kolstad, Barnes. *A Critical Study of the Manufacture of Soft Mud Building Brick.*
Nutt, John. *Newburgh.*
Ries, Heinrich. *Clay Industries of New York State.*
Rosenwaike, Ira. *Population History of New York City.*
Searle, Alfred. *Modern Brickmaking.*
Seger, Hermann. *Seger's Collected Works.*
Sylvester, Nathaniel B. *History of Ulster County.*
Ulster County Clerk's Office—Deeds and real estate transaction information.
U.S. Bureau of the Census. *Historical Statistics* (Cost of Living Index).
U.S. Bureau of the Census. *Products of Industry.*
U.S. Patent Office. (See Appendix).
West, Richard. *Functions of Coal in the Manufacture of Hudson River Soft Mud Brick.*
Weatherby, Hazel C. *Nearly 300 Years of Moise Quantin and Descendants in America.*

Addendum to page 47, "Drying": An edging device turned bricks from their initially flat position on the ground to resting on one edge for drying of the surface that had been on the bottom. (Courtesy of John Pawloski, Director, The Connecticut Museum of Mining and Mineral Science)

CHAPTER 3
Zenith of an Industry and Irresistible Forces of Change—1897-1920

FOR THE HUDSON RIVER BRICK INDUSTRY as a whole, this period of time is both zenith and sudden decline, all in an extraordinarily short period of time. The turn of the century was also a period of accelerating technological change. The Hutton Company, given the lack of experience of its new management, was lucky to pull through this era.

With the 1897 death of William Hutton, the company went from sole, unincorporated ownership to fully incorporated ownership by six family stockholders, with elected family officers and directors. Only one of these, George Hutton, had any known previous experience with the brick business, for only seven years, and probably with a minimum knowledge of actual manufacturing operations. He became secretary-treasurer, his younger brother Robert Kerr Hutton became president and his mother, Jane, became vice president. They all arrived at this juncture in a troubled circumstance.

The new officers and directors, all family members, were immediately faced with the results of the depression which the banking crisis of 1893 created (causing a drastic nationwide shutdown of construction). The depression caused a thirty-percent wholesale brick price drop to $3.75 per thousand for Upriver bricks—the lowest price since 1879—and it stayed close to the $5.00 break-even price for three years before it grudgingly went back up in 1898, allowing a modest profit. A price difference for Upriver bricks of as much as $1.00 per thousand lower than Haverstraw bricks was troubling, and lower spreads were usual.

A depression could go away, but permanently intractable, industrywide problems were rapidly accumulating. In 1898, the new Hutton Company directors were required to individually contribute substantial amounts of personal capital to the company. A bond in the amount of sixty thousand dollars ($1,320,000 in the year 2000) was also sold, using the company's property as collateral. The reason for the bond is not known, insofar as there is no indication of resulting capital investment. It is assumed that its purpose was for provision of liquidity—cash—for operations. This was a rough start for the new officers and directors. Recognition of the chronic condition of oversupply in the market and its effect on brick prices—the intractable problems—resulted in the formation of the Haverstraw Bay Brickmakers Association in 1893, followed by industrywide meetings in the following year at a redoubtable institution of the contemporary Newburgh life, the Palatine Hotel. The severe nationwide depression brought into sudden focus the results of the overproduction that had gradually accumulated.

An Inside View of a Business Crisis

Running a revealing series of articles concerning this subject in 1893-1894, the *Rockland Messenger* described in considerable detail the angst and travails of the manufacturers at a turning point in the fortunes of the Hudson River brick business. Singular enough was the willingness of the Haverstraw manufacturers, in the spring of 1894, to reveal their very serious concerns and the inner workings of their efforts to change the deteriorating condition of the business. The theme of those efforts was clearly expressed from the beginning: "In view of the low price of brick at this time, the undersigned deem it advisable to take some action whereby they may endeavor to increase the price. . . ." From the start, a significant number of signatories to the idea of collective action were enlisted, with various degrees of persistence thereafter. Plans included the formation of a permanent organization. At least two of those early enlistees, namely the Pecks and Excelsior, went their own way because of their large size and their "willingness to be opportunistic about prices."

Market Instability

The brokers ('commission men') in the city were acting against the new Haverstraw Brickmakers Association. The commission men were selling bricks from outside of the association—mainly Upriver bricks—at a lower price. Therefore, an immediate effort was made to include the Upriver manufacturers in the association. The price spread between Upriver and Downriver bricks generally varied from five to twenty-five percent. Haverstraw manufacturers were very sure of the qualitative superiority of their product—and the makers of 'Haverstraw hard' bricks gave the market some kind of justification to that claim when they expressed calculated wonderment about Upriver bricks "so heavy you wonder how they could bring that price." Anxiety was plainly expressed over the possibility that the largest manufacturers might stay outside of the association in order to drive small manufacturers out of business.

At a meeting at the Palatine Hotel in Newburgh in early 1894, assembled manufacturers were presented with three alternatives: to agree on a mutually beneficial plan of manufacturing and marketing, to join a trust controlled by outsiders, or to continue along in the manner in which they already had harsh experience. The meeting produced agreement to limit production by means of delaying the starting time until May 20, using only half of their productive capacity after that date. Some favored the idea of a trust or a syndicate, with all yards under central management by a manager or a board of managers. Naysayers included the Brockway interests who flatly stated that they could produce at one dollar per thousand less than the others (twenty percent lower at the current prices) and that they wanted no part of any cooperative efforts. There was also speculation about the likelihood that, "those controlling the syndicate will receive their big salaries and stockholders will wonder where their stock is quoted at." Also proffered was the possibility of the syndicate having to send brick to the market at very low prices for two or three years in order to crush the manufacturers who were outside of the syndicate, "to get the upper hand of the business." Those events would be followed by improvement of the plants and "a plan arranged whereby the stock will go down for the purpose of enabling strangers to buy up and finally own the yards, clay banks and stock, etc." The opinion was expressed that the manufacturers should get busy with the formation of their own trust before an 'outsider' got in ahead of them. An outsider did indeed appear, in the person of Charles R. Flint of New York, "who controls more capital than any other man in the city. . . ." This person was "satisfied that if the brick business can be combined, larger dividends can be obtained. . . ."

Proposed Consolidation

Charles Flint had a clear understanding of the need for consolidation. Joiners would take preferred stock as payment for their facilities and those who thereafter elected to leave the syndicate would be paid in cash for their plants. He wanted it understood that the manufacturers, "must not expect a big bank handed over to them at the end of the year. . .but could expect a fair dividend." Reactions from manufacturers included speculation: "it seems very strange if strangers step in and are willing to put their money on businesses in which home manufacturers claim they are losing money. . . ." and generally not being favorable about 'strangers' running their businesses. The New York and New Jersey manufacturers made other efforts to form a trust, but no conclusive action was taken. The hand wringing continued: "The brick business is being run without a head and brick are sold without discretion or consulting with manufacturers. . . . There are enough brick to last until August and if we commence May first there will be no help for us. It will flood the market before the first of July. There is no system in the business." Shutting down the brick business for one year was strongly recommended. That is exactly what happened at the big Aldridge and Ramsdell (Dennings Point) yards at Dutchess Junction and Fishkill Landing. A Coeymans manufacturer was willing to forego his tenants' rent payments if they ceased operations for a year. A manufacturer at Dutchess Junction (just south of Beacon) added that he had run his yards at one third of capacity for two years without making money, while making substantial investment in his plant.

The predicament of pricing in the New York market was also discussed, recognizing that the commission men had tried to favor the manufacturers. The scenario was stated as follows: insofar as their income was derived from the volume of brick sold, the commission men were forced to sell regardless of price. Knowledgeable buyers used this weakness as a means to force prices down, with nothing to oppose this. "There was no force to hold the price up, other than the human voice and a self which money could alone break down." It was understood that changing the basis of the commissions, to be based upon the total value of the sale, would create a better condition for the manufacturers.

By the mid-summer of 1894, the business was showing obvious signs of critical strain. Some employees plainly understood this, and they offered constructive measures for sharing the pain while continuing operations, as at the Corse yard in Glasco. Those gentlemen contracted with the owner to operate the yard for the rest of the season "on shares." Other solutions, lacking the same finesse, included delay of the start of manufacturing until September, to outright suspension of operations, and to cut salaries and wages. In August, bricks were selling for four dollars per thousand—a disaster. In December of that year, improved market conditions had advanced prices by $1.00 to $1.25 per thousand—but with a surprising change in the relationship of commission men and buyers. The buyers were dealing directly with the manufacturers, thereby circumventing the commission men. This foreshadowed the situation provided in the next decade by the John B. Rose Commission House's operations, described later in this chapter, as well as the Greater New York Brick Company.

The Possibility of Consolidating

Heinrich Ries noted the inability of Hudson River manufacturers to consolidate. In his 1909 history of American clay-working industries, Ries stated that consequential consolidations of brick companies had occurred in Baltimore, with the 1899 formation of the Baltimore Brick Company; in Boston, with the 1900 creation of the New England Brick Company, as well as in Chicago, with the Illinois Brick Company, also in 1900. It was noted that such consolidations were made in the interest of maintaining prices 'at a reasonable figure'—and such consolidations must surely have occurred as a result of very similar circumstances of overproduction that faced the Hudson River manufacturers. The Illinois Brick Company persisted into the mid-twentieth century as a very large maker of common brick whose low prices incurred the animosity of much higher-priced face brick manufacturers in the Chicago region after the middle of that century.

Adding to the manufacturers' dilemma was a long strike in 1898 that affected the Downriver yards but left the Upriver yards in operation. Downriver lessees were at particular risk, and the manufacturers regarded this strike as a genuine threat to the existence of the Hudson River brick industry. The silver lining to this dark cloud was that it did provide needed curtailment of production.

Failed Consolidations

Awareness of the industry's problems seemed to have also gone out into the business world at large. Charles Flint's proposal to the manufacturers of 1894 has already been noted. Some individuals sense an opportunity in such situations. The appearance of Oakley Thorne is some kind of evidence of this possibility. Thorne was president of the National Switch and Signal Company of Easton, Pennsylvania, and also had a sizeable estate near Millbrook, New York. The plight of the Hudson River brick industry attracted his attention to the extent that he solicited the manufacturers with a letter (dated January 1899) that proposed a corporation, owned by Thorne, that would purchase seven hundred million brick per year at a contracted rate, reasonable for the manufacturers, thereby effectively controlling the price of bricks in the metropolitan market. The manufacturers would be protected from the market's vicissitudes. Of course, the idea of cornering markets was very fashionable at the time—and also was beginning to receive the increasing federal government scrutiny that had already begun with the Sherman Act of 1890.

For the advancement of his purposes, Thorne formed the American Brick Company in 1899, on Staten Island. The local manufacturers didn't cooperate on this one either. After all, Thorne was also an outsider to Hudson River brick circles. However, to the officers, stockholders, and directors of the Hutton Company, Oakley Thorne was an angel from heaven. Incredibly, he proposed to get them out of the very troublesome brick business by purchasing the company—which he did, in November 1899, in the name of his American Brick Company. The company made at least one additional known brickyard purchase at Beacon, in the same year, and several others had reputedly been purchased as well. The purchase price for the Hutton Company was two hundred thousand dollars, with the directors receiving ninety thousand dollars 'on account of the purchase price.' That is not the end of this story. In March of 1900, four months later, the company was sold back to the Hutton Company for one hundred sixty thousand dollars, a profit of twenty percent to the Hutton Company owners. A similar sale of the Beacon yard was made back to its original owners, at similarly advantageous terms to them. The Hutton Company directors at least recovered their personal loans to the company—and they were back in the brick business whether they liked it or not.

The above is a recitation of facts taken from Hutton Company directors' minutes and from two deeds of sale (from Hutton Company to the American Brick Company and from Thorne to the Hutton Company). The company thereafter proceeded with equipment purchases, buying barges from the Hildebrandt shipyard (located on the Rondout Creek at Connelly), and new brick machines. The Hildebrandt yard furnished barges and drydock services to the Hutton Company for decades to come.

Another redoubtable institution, Kingston's Eagle Hotel, served as a rallying point for the Ulster County manufacturers. The Eichler Hotel served the same function. The latter had a handsome small bar with a large circa 1900 mural painting on the back bar wall featuring a very pleasant-looking Levantine damsel (fully clothed) among a cascade of exotic flowers, all of this being in place as late as the early 1970s, but tragically demolished thereafter, thereby also eliminating the last ten-cent glass of beer in existence (small as that glass was).

Hutton Company directors' minutes note 1906 as a year that featured a fifteen-day strike, a pay raise, and tactics for dealing with the corrosive effects of a new saloon on North Street. It was thought the saloon had an unwholesome proximity to the brickyard. The strike indicated that news of the sudden upward 1905-1906 spike in brick prices had gotten around. The proposed solution to the North Street saloon seemed to have been the encouragement of a more distantly located competitor saloon out on Kingston Point—a more reasonable compromise rather than the axe-wielding, saloon-smashing kind of high-handed, moralistic temperance that characterized some of the contemporary opposition to such beverage purveyors.

Ceramic Technology—circa 1900

Recognition should be made of the state of ceramic technology in the early twentieth century, with scientific knowledge much more advanced than might be expected. Some of the names of principal contributors to this background have already been mentioned, with the indispensable Heinrich Ries and the authoritative German ceramic scientist Hermann Seger, both of whom showed a broad range of knowledge of various elements affecting the product, as well as specifics of kiln firing and drying technology. Ries was a professor of economic geology at Cornell University who wrote on a wide range of subjects including several that most specifically related to Hudson River clays and brickmaking practices—as this writing shows in its extensive citations of Ries's work. One of Seger's greatest contributions to the ceramic industry was the invention of 'pyrometric cones,' a device for measuring the high temperatures within kilns (a grouping of finger-like projections, on a common base, that melted, one at a time, at progressively higher temperatures). They were cheap, extraordinarily effective, and are in use down to this day. The English authority Arthur Brown wrote extensively on drying (1902). He included factors that affected the process and described the most advanced methods in this area of the manufacturing process. Other well-informed technical writings appeared in the definitive trade publications *Brick* and *Clay Record* (combined into *Brick and Clay Record* after 1910).

Significant advances were made in Hudson River brickmaking technology, including the gradual abandonment of steam power in favor of electric power. This subject received considerable attention in the brick industry journals during the first decade of the twentieth century. Electric power could be applied to a large number of specific tasks including work done by power shovels, tramways and railways, pug mills, brick machines, conveyors, blowers, etc. Direct-drive electric motors eliminated transmission losses from belt slippage (thirty to sixty percent of engine output) and the friction from the many bearings needed to support lengthy 'power line' shafts, as well as breakdowns of that shaft system that provided power to many operations along its length from a central steam engine source.

Belt-driven power still had a lengthy future. New belt technology provided the mechanically glamorous, slip-proof metal link belts that were also vulnerable to wear from grit and dust (though advertised in the brick industry trade journals before 1920). The now-familiar vee belt, of composite material, rapidly became the dominant type of belt drive during the 1920s, these too being featured in trade journal advertisements. Depreciation of electric motors was estimated to be one half the rate for steam engines. Electric power units could be individually provided for each piece of equipment, providing far greater flexibility in allowing shutdown of limited portions of the production system when required, as well as allowing precise allocations of power for efficiency of electricity usage.

One remarkable tribute to steam power had to be the operational persistence, until 1949, of a late-model steam-powered brick machine, that had a very respectable production of eighty-five thousand bricks per day, at the Powell and Minnock plant. That plant was among the top six plants during the post-World War II period and was the sole survivor into the third millennium.

More Innovations

A new name would be added to the list of illustrious Downriver innovators that already included James Wood and Richard Ver Valen. This new entry was David Strickland. His inventions were widely used in the American brick industry. The first was a brickmaking machinery patent in 1909 that was assigned to the American Clay Machinery Company of Columbus, Ohio. The highly successful 'steam dryer' system of 1912 followed. The steam dryer used steam piped from a boiler to a room that had racks made of pipes which supported the brick pallets, with these pipe racks actually filled with steam—thereby providing both support for the bricks and a conduit for steam heat. The radiant heat thereby provided was an excellent means of drying the wet clay. Heat was conducted from the hot pipe surface through the steel pallet to the damp brick as an additional boost to drying. The earliest usage of a steam drying system on the Hudson River appears to have been at the Ferrier and Goldin yard in Catskill, in 1890, but it was short-lived.

Already in use elsewhere in the United States during the very early twentieth century was the concept of kiln heat reclamation for the drying process, using blowers for hot air circulation. This was possible with kiln types that lended themselves to the ducting of remainder heat from kiln firing (heat from cooling bricks otherwise expelled into the atmosphere), such as the Hoffmann and beehive kilns. The reclaimed heat was introduced directly into tunnel chambers that contained rail-mounted carts bearing the newly-formed bricks. However, the scove kiln's large firing area absolutely precluded any such heat reclamation, making Strickland's steam heat innovation an ideal adjunct for the scove kiln yards. The steam heat system did require installation of a separate low-pressure boiler plant and added the cost of fuel for steam generation to manufacturing costs.

'Tunnel dryers'—heated either by hot air or by steam—became an accepted technology in the American brick industry, with the Cary yard in Columbia County having such an installation in 1903. Recall that a hot air drying system had been used at the Van Vliet yard on the Fishkill Creek during the late 1850s, probably using wood as a fuel for air heating.

New Additions to Kiln-Firing Technology

The use of oil as a kiln fuel had already made its first appearance in 1898, according to C. E. Hall, at Haverstraw's Garner yard, and gradually spreading through the valley's other yards. Hall did not seem to know about the use of oil for kiln fuel at S. E. Coykendalls' Port Ewen yard, cited in Ries's writing of 1895.[1] That installation necessarily preceded the publication date—thereby not only putting it in serious contention as the first to switch to oil but doing so at a time when most yards were switching from wood to coal. Another account[2] states that the Ferrier and Goldin yard at Catskill was the first to use (crude) oil as a kiln fuel, as early as 1886. Heinrich Ries, in his writing of 1900[3] stated that oil was the cheapest fuel of all, with the only available cost information stating a royalty of $1.60 per burner (an unsatisfactory rate that doesn't account for volume of oil used).

Combining soft coal and oil for kiln firing became a common practice, allowing a considerable reduction in costs—and this combination was used in scove kilns well past the mid-twentieth century.

Another significant improvement, also originating in 1904, was the introduction of ducted forced air into the kiln arches in order to insure adequate draft and oxygen supply for the combustion of the coal contained within the bricks. This could be used with oil, coal, and wood fuels, and was an important contribution toward improvement of scove kiln performance.

Figures 44 above and 45, below. Clay Banks. Presumably at Rose Company, Roseton, late-nineteenth century. (Courtesy of New York State Archives)

Figure 46. Clay Bank Railroad. Location unknown, late-nineteenth century. (Courtesy of New York State Archives)

A Brief Survey of Manufacturing Technology at the Zenith of the Industry

A notable account of the brick industry in 1905[4] describes the manufacturing practices of fourteen yards in considerable detail, thereby rendering a view of an industry that was operating with many methods that had not appreciably changed in decades, as well as making use of progressive technology. Indeed, the atmospheric drying yards described therein had been around for several millennia.

It must also be said that the Hudson River industry was far from unique in the retention of open drying yards. Such practice was widespread in the United States at the time. At that moment, the largest yard on the river, John B. Rose's company, was using thirty-five thousand cords of wood (no, that quantity is not a typographical error) per year for kiln fuel. A special contract with the nearby E. H. Harriman estate supplied the wood. Rose genuinely believed that wood firing of kilns produced a better brick. Rose was also using archaic sod pits for mixing, believing that properly filled pits were better than the circular tempering wheel pit—and he also used open drying yards.

John B. Rose

In 1908, Rose became precocious in other parts of his operations, installing an electric railroad, with five locomotives, to his clay banks, where clay cars were loaded by three steam shovels. John Rose's own manufacturing operations were four miles to the north of Newburgh and capable of an annual production of between forty-two million and one hundred million bricks, allowing him to claim the title of being the world's largest producer of common brick. Rose took over his father's business at age twenty-two, only two months after graduating from Yale College in 1897. His family had originated in the Town of Esopus, just south of Kingston. Other members of the Rose family owned a yard to the north of Kingston, at Ulster Landing.[5]

Figure 47. John B. Rose.

John B. Rose became a presidential elector for Theodore Roosevelt in 1904 and was elected to the New York State Senate in 1906. At the turn of the century, the Rose Company's operations were so consequential that an entire community for brickyard workers, named Roseton, was created. This center of brickmaking included two other adjacent big manufacturers—the Jova Company and the Arrow yard (the clay bed depth at Roseton varied between 180 and 240 feet, a depth unmatched at any other location on the River).

By 1905, the Arrow yard fully electrified its brick machine power, a notably early conversion from steam power, while also persisting with open drying yards. A handful of the manufacturers were supplementing the usual circular tempering pit and sod pits with a separate pug mill in order to obtain superior mixing, and one yard using the pug mill alone. This showed commendable critical analysis of manufacturing processes, insofar as the mixing boxes on the brick machines were probably only doing a marginal job of mixing the clay and sand contents. As previously noted, a limited number of Haverstraw yards used covered drying yards or palette yards, those manufacturers being among the largest and most fully capitalized.

The scove kiln remained the great common denominator of all the yards, and coal was the predominant fuel at the turn of the twentieth century. As a portent of momentous changes that was decades away, a full-page cover advertisement in a 1918 issue of *Brick and Clay Record* announced the availability of "American Dressler Tunnel Kilns." This was a radically new type of kiln, which is extensively discussed in following chapters.

The Mortal Threat of New Construction Material

The first decade of the twentieth century also marked the fully assertive entry of the common brick's greatest competitor—concrete made with the rapid-setting Portland cement. Heretofore a minor but growing presence, it gave notice in an article in the journal *Brick* of 1905, with respect to the increasing volume of concrete block production: "You all know how rapidly the use of cement has spread. . . . The amount of Portland cement used in this country doubles once in four years." The same issue did concede that concrete was cheaper—and also noted concrete bridges that were exhibiting spalling and cracking. Also very much at issue was the question of the ability of concrete to withstand fire. Regarding the performance of brick during the great Baltimore conflagration of 1904, the Committee on Fire-Resistant Construction of the Underwriter's Association stated: "Ordinary well-burned brick of good quality is the most satisfactory fire-resisting material now used in building construction."

The Baltimore and San Francisco fires occasioned the testing of panels for fire resistance by the U.S. Geological Survey. Those made of brick were the least affected, with concrete panels being badly pitted. The 1905 *Brick* article also cited systematic materials testing by the New York City Bureau of Buildings—for compressive strength, fire resistance, water absorption, and multicycle freeze-thaw tests—that confirmed the very durable qualities of common bricks. Needless to say, the brick industry widely touted all that kind of hard evidence in favor of brick.

Fifty years later, the latter two types of tests (water absorption and freeze-thaw) were a very principal and unjustifiably fatal subject to what was left of the Hudson River industry. At the beginning of the twentieth century, brick was still holding its own as a principal building material. Another negative, just as damaging to the brick industry as the increasing use of concrete, asserted itself during the next decade. However, the growth in use of Portland cement was prodigious, at a rate of eighteen percent per year during the first half of the twentieth century—a rate of increase in usage unsurpassed by any other material during that time with the exception of aluminum. By the end of the first decade of the century, the annual use of Portland cement had increased by a multiple of eight times the consumption at the beginning of that decade (from ten million barrels to eighty million barrels).

Population Trends

The other variable, the city's population trend, was powerfully positive during the first decade of the twentieth century, with a thirty-eight percent increase of 1.33 million people. However, the following decade—1910 to 1920—saw an increase of only eighteen percent, adding 850,000 people. This is still a large number, but a significant drop from the previous decade. The years of the great New York City population increases were over. In fifty years, 1870 to 1920, the city grew from 1.3 million to 5.6 million, while consuming vast amounts of bricks. Notably, growth began shifting away from the original growth centers of Manhattan and Brooklyn—a process that continued for decades to come. The newly developed subway lines and other commuter rail systems hastened this. The subway line commenced operation in 1904 northward into Harlem (causing a wave of real estate speculation that resulted in large scale overbuilding of apartments, with newly constructed buildings entirely unoccupied), and expanded rapidly into the Bronx and Queens after 1913.

In 1910, Manhattan hit a population peak (2,330,000) after a twenty-six percent increase in the the first decade of the 1900s—followed by a 20 percent population loss by 1930. The completion of the first subway tunnels to Brooklyn and the Bronx by 1908 began the process of depopulation on Manhattan Island. Brooklyn's growth rate was cut in half, from 40 percent in the first decade of the 1900s to 23 percent in the second decade. The Bronx had a 115 percent growth rate in the first decade of the 1900s, followed by hefty increases of 70 percent for the two following decades. This was the first big increase outside of the old growth centers of Manhattan and Brooklyn. Sizable expansion in Queens came during the 1920s. A trend was being established that most certainly affected the Hudson River brick industry. It had heretofore relied on a market that was very proximal to cheap water transportation, and was centered around multi-story, densely-sited urban development that required good fire resistance.

Too Many Brickmakers

Throughout the first decade of the 1900s, a system of representation of the manufacturers, marketing, and sales had become well established, if overlapping and troublesome. The weak link in this was the manufacturers themselves. Additional tentative efforts were made to establish a cooperative association for the control of overproduction. The real need was for a reduction in the sheer number of manufacturers that included significant numbers of lessees. Effective action on such a matter was out of the question for any voluntary association of manufacturers. Owner-lessors were themselves guilty of exacerbating the oversupply problem in their unwillingness to curb their leasing activities. Their shortsightedness in this regard only harmed them in obtaining an adequate price for their own bricks. Leasing had become a habit that the clay landowners were unwilling to break. Bear in mind the initial failure of the Haverstraw Brickmakers Association of 1893 in obtaining effective cooperation from the manufacturers. The multiplicity of such organizations makes it difficult to present a clear picture of the turn-of-the century manufacturers' associations.

More Attempts at Consolidation

According to one contemporary account, the manufacturers formed a jointly held corporation called the Consolidated Brick Company, with no resulting solutions to common problems after two years of its existence. An 1898 entry in William Hutton's estate audit[6] that shows five shares of Consolidated Brick Company stock verifies the company's existence. A subsequent entity was formed in 1902, the Manufacturers Consolidated Brick Company. This also expired from mutual suspicions concerning the cornering of portions of the market by some of its members, while using the good offices of the organization. Additionally, the journal Brick states the existence of the Hudson River Brick Manufacturers' Association, founded in 1902, comprised of one hundred and twenty-six firms. This association established district commissions that filed weekly reports to the association regarding prices, labor conditions, yard conditions, etc.—all the signs of a functioning service organization. The district commissions also sponsored brick-makers' clubs that met almost daily for discussion of sales and marketing problems.

The association revealed its weakness as a means of wresting control of prices in the plainly expressed dissatisfaction of the members regarding the association's failure to modify the ability of the Mason Materials Dealers Association to fix the prices obtained by the commission men—it was the dealers who controlled the prices. There was alleged involvement of Brick Association members serving as members of the board of directors and executive committee of the Masonry Materials Dealers Association, although it is not clear in what manner this involvement affected prices. They certainly wouldn't use their influence to lower prices on their own bricks.

Generally, manufacturing operations persistently bore the stamp of oversupply. It was agreed that supply would be reduced by mutually agreed starting dates postponed from March into mid-May. Such scheduling significantly reduced the customary work year of eight months—a critical hardship for workers, which threatened their existence in the industry. Such agreements were subject to avoidance by individual manufacturers.

Decisive Actions

A principal outcome of the manufacturers' lack of decisive action was the appearance, in 1902, of the John B. Rose Commission House, owned and operated by the largest brick manufacturer on the river. The enormous brick market at the west end of 52nd Street was the principal feature of the Commission House. There was additional marketing power in the ownership of four of the largest brick retailers in the city. John B. Rose was unquestionably a young man (twenty-seven years old at the time of the founding of the Commission House) of vision and decisiveness, however flawed the final outcome of his undertakings. The Commission House was the direct result of Rose's exasperation with the manufacturers' inability to cooperate in resolving the common problems of marketing. In one year, the Commission House could move six hundred million bricks, a significant percentage of all brick sold on the New York market. This brokerage represented fifty manufacturers, including eleven from outside the Hudson River region—yards from New Jersey, Delaware, and Connecticut. Deliveries from the yards were made by 130 barges and ten sail-powered vessels, as well as two hundred horse delivery teams within the city. The brokerage also provided regular reports to its manufacturer-customers on market and inventory conditions. Forty barges could be berthed at one time at the West 52nd Street docks on the Hudson River (later the location of piers for the great transatlantic liners), awaiting inspection of their cargoes by prospective buyers. The Commission House was also immediately benefiting from the great surge in demand that crested from 1905 to 1906.

The manufacturers made one final effort to organize for the purpose of price stabilization, in the Greater New York Brick Company (the Greater New York) of

1911. One model that caught the attention of the organizers was another Chicago corporation created for purposes of achieving price stability, the Chicago Brick Company, in addition to the aforementioned Illinois Brick Company of 1900, even though both of those companies seem to have been true consolidations of other companies—something beyond the capability or willingness of the Hudson River manufacturers. By 1911, the Illinois Brick Company had become the largest brickmaker in the world, with nine plants in operation and several that were not in production (in all probability inefficient producers that had been bought up at low cost in the parent company's consolidation drive). Those Chicago consolidations were in response to competitive pressures quite similar to those in New York City. There was only a handful of multiple plant owners on the Hudson River, such as William Lahey with his yards at New Windsor, Fishkill Landing, and East Windsor Hill, Connecticut.

The Greater New York was not the result of cooperative effort among the manufacturers, but a hybrid organization resulting from the initiative of a handful of New York brick brokers—the commission men. Even in the last century, the Cordts and Hutton Company had an agent in the city, early in the 1870s, by the name of Robert Main (born in Crookham, Cumberland County, England, in 1850 and immigrating to America in 1864). Main was captain of a freight boat by age twenty after which he was employed by Cordts and Hutton. In 1880, he established his own brickyard just north of the Terry yard (Kingston) and subsequently at Stockport (Columbia County). The Mains were also brick manufacturers at Glasco in the early twentieth century. William Barnes, John McNamara, and Richard Morrissey were prominent names among the early twentieth-century commission men. Barnes was originally in the brick business at Grassy Point (the XXX brand) with the Farleys. A son of the latter family, James, became a nominee for the U.S. presidency and then served as postmaster general during Franklin Roosevelt's administration. Jim Farley then became president of General Builders, a large materials dealer in the city that also purchased the Dennings Point plant (opposite Newburgh) in 1946. Farley was famed for his legendary ability to unfailingly associate an enormous number of peoples' names with their faces, priding himself on instant recall.

The Morrisseys were also in the brick business on the Minisceongo Creek (at Grassy Point). Barnes, McNamara, and Morrissey, by World War I, became a principal brick brokerage firm in the city and remained in business into the middle of the century, with the Hutton Company as a significant account. By the middle of the first decade of the century, the commission men were already working closely with the manufacturers, at times furnishing financing as well. Many of those brokers were themselves originally from Haverstraw and Grassy Point and on good personal terms with many of the brickmakers. Nevertheless, the objective was *to produce the largest number of brick at the lowest cost.*[7] That was the market-imposed condition for the manufacture of common bricks for the New York City market—and the negative effects of this inexorable pressure on the industry's profits, and consequently upon investment in new technology, plant and equipment, wages, and working conditions cannot be overemphasized. By 1911, most of the manufacturers were forced to recognize the advantages of being under the corporate wing of the Greater New York. Its leadership was well versed in the ways of the New York market. The new corporation had modest expectations—to move the price from the 1911 price of $5.78 per thousand to $6.12—a price that would allow operating without a loss. The break-even price had moved to $6.00 per thousand. But there was little room for real profitability, insofar as there were significant brick industries in Connecticut and New Jersey waiting for prices to rise to levels that would trigger their entry into the New York market—$6.50 in the case of Connecticut.

John B. Rose was elected president of the Greater New York, with the intent of terminating the Rose Commission House. The new corporation represented eighty percent of the entire Hudson River production. Such a degree of cooperation undoubtedly resulted from the chaotic condition of the market at that moment—cooperation born of raw anxiety. There were two essential purposes for The Greater New York—the principal of these was to obtain a satisfactory price for all the members' production. In addition, this new corporation sought to act as a central purchasing agent for materials needed by the members, in order to obtain economies of buying in large quantities. Such materials included coal and sand. The latter was stripped from the clay bank overburden at many of the yards, thereby leaving some manufacturers without a source of usable on-site sand.

The Greater New York did successfully upset the conditions for market control that had been achieved by the building materials dealers and the other commission men who were outside of that organization—and they were not philosophical about it. Such a challenge to their own control brought on a petition to the New York State Attorney General, alleging the existence of an anticompetitive trust. The Greater New York people were depicted as rapacious corsairs, reputedly amassing the might of ninety percent of the industry's manufacturers into a monopolistic juggernaut. How was the Greater New York organization obtaining cooperation among its customarily unruly members? It was alleged that much of the leadership was closely connected to the principal banking houses in the Hudson Valley, putting them in a position to coerce less well-positioned manufacturers by withholding critical financing from those banks. The Attorney General ruled that the evidence against the Greater New York was insubstantial and the case was dropped.

The Zenith and Collapse of the Hudson River Brick Industry

In 1905 and 1906, the Hudson River brick industry arrived at the greatest level of production in its existence. Prices in those years were only exceeded by a sharp spike in 1923, hitting a momentary top of twelve dollars per thousand in late 1905 and early 1906, with an average of eight to ten dollars. At that moment, annual production was 1.3 billion bricks manufactured by 131 Hudson River brickyards. Such quantity is almost beyond comprehension, requiring some kind of exemplary equivalents. One point three billion bricks would provide a thirty-two-mile, solid brick length of the Great Wall of China (only faced with brick and filled with rubble) or a three-hundred-mile long (Albany to Buffalo) wall four stories high (each story being ten feet for a total height of forty feet) and twelve inches thick (required by code and braced by floor framing), with no window or door openings. Driving time to traverse the length of that wall would be five and one half hours at the state speed limit of fifty-five miles per hour. The Hudson River brick industry was at the very top of the very pinnacle of its existence and at the zenith of a roaring market. But, also in the manner of that same industry, there immediately followed a precipitate drop in production to 800 million bricks in 1908 and the price to $5.10. A very high production volume in 1909, 1.2 billion, which was exceeded only by 1906, followed this. However, the price differences between those two years indicates a deaf ear for the market's message. In 1906, bricks were at an average of $8.00 to $10.00 and in 1909, they were at an average of $5.28, with the latter plainly signaling a glut of brick.

One journal, *Clay Record*, showed 1907 prices between $4.50 and $5.50, with the manufacturers' cost to put the brick on the New York City dock at $6.00 per thousand. Such conditions brought on the usual big meeting of manufacturers (one hundred representatives) at Newburgh's crusty Palatine Hotel with the usual inability to produce any workable recommendations. To the individual brickmaker like the Hutton Company, such volatility caused a production cut. They accomplished the cut by using only four out of their twelve brick machines—and the Hutton Company was among the top six manufacturers. A ruinously low yard price of $4.54 was reported in 1910, with almost a third of the year's production remaining as unsold inventory and, by 1911, twenty-five of the manufacturers had left the business (a reduction of almost twenty percent). This was the first blood-letting of the ranks of the manufacturers. The Greater New York company was formed the following year. It had a nearly impossible situation to deal with, and the attainment of a break-even price seemed like heaven itself.

Ulster County Takes Production Leadership

Such low prices at this moment were the result of a multiyear downturn in construction activity. The cause was the national financial panic of 1907 (a creature of Wall Street so critically severe in its moment of crisis for the American financial system that it lead to the establishment of the Federal Reserve system in 1913). Cement prices were also down by thirty-five percent and structural steel prices down by thirty-two percent. Such price levels in a generally weakened construction market persisted, with short intervals of improvement, for eight years. That 1907 financial panic was the fatal cut that brought on the collapse of a large part of the Hudson River brick industry at the greatest moment of its existence—and it was in that same year that Ulster County took over the position of being the

Figure 48. Ulster County Brickyards

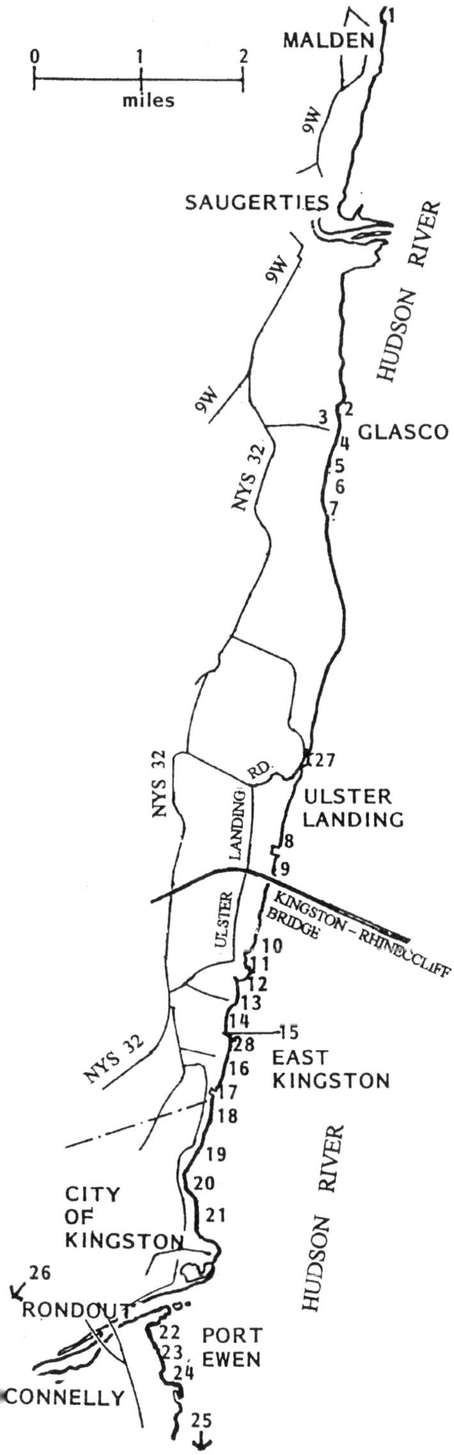

KNOWN LOCATIONS	EXACT LOCATIONS UNCERTAIN
MALDEN	
1 Staples 1903-1958[7]	Bishop[7]
	Cooney & Farrell[6]
GLASCO	
2 Mayone	Alonzo Rose - 1890s[6] *
3 Robert Lent-1850s[2,8]	Gaston Wilbur -1860s[2]
4 Henry Corse-1870s[4]	I. Dunn & Co.[5]
5,6 Washburn Brothers-1867-1849	Empire Brick & Supply Co.[7]
(upper & lower yards)	Alpha Brick & Holding Co.[7,8]
7 Main 1880s;[5] Robert Main[8]	J. Porter[6] *
	John Overbaugh - 1840s[1]
	Overbaugh & Maginnis - 1840s[1]
	F. N. Van Deusen[6] *
	William Hale - 1840s[1]
	Gardner & Goldsmith[7]
	Van Duzen & Terry[5]
	John Maginnis 1840s[1]
	C. H. Littlefield[6] **
	D. C. Overbaugh[6] **
	* Immediately south of Washburn
	** 1 mi. south of Washburn
ULSTER LANDING & TOWN OF ULSTER	
8 Goldrick - 1900s[7,8]	Daniel Weed[2]
9 Rose Bros.[8]	Weed & Sutton[2]
10 Terry Bros.	John Shutero[2]
11 Goldrick (lower yard)	Palmer Gerrit[2]
13 U. F. & J. T. Washburn-1870s	Thomas Beekman[2]
12 Brophy	Ulster Landing Brick Co.[7]
14 Streeter & Hendricks,[5] Hendricks[7,8]	Lynch Bros.[7]
27 Hanrahan Brick & Ice Co.[8] (Dwyer)	
EAST KINGSTON	
15 Terry Bros.	
16 Brigham - 1891-1950s	
17 Charles Shultz[4] & Estate of C. A. Shultz[6,8] 1876-1916; Successor owner to 1940	
28 Manchester[14]	
KINGSTON AND PORT EWEN	
18 Nathan Nickerson,[4] Staples,[4]	
East Kingston Brick Co.	D. Manchester[11]
19 Fredericks,[5,7] R. Main,[12]	Van Deusen[12]
Kingston Brick & Ice Co.[8]	G. Kerr[5]
(Dwyer)	E. Gardner[13]
20 Terry Brothers[2] 1850-1946; successor owner to 1956	
21 Cordts & Hutton 1865-1890;[9] The Hutton Co. 1890-1965;[10] successor owners Jova Mfg. Co. and Staples to 1980	
22 Turner 1890s (sold to Gurney)	
23 S. E. Coykendall[5,6]	
24 Jacob Kline[5,6] (sold to Bishop[8])	
OTHER LOCATIONS	
25 Schleede (Ulster Park)[7,8]	
26 Lowe (New Paltz)[8]	

NOTES – Sources Listing Manufacturers (with dates of listing); numbered sites are from on-site identification.

1 1850 U.S. Census – Products of Industries
2 1860 " "
3 1870 " "
4 1880 " "
5 *List of Manufacturers Tributary to the NY Market* 1890
6 Ries 1895
7 deNoyelles – Brick Mfrs. Supplying the NY Metropolitan Area, 1910
8 NYS Dept. of Labor 1913
9 Deed to Cordts and Hutton 1865
10 *Kingston City Directory* 1890; (11) 1892; (12) 1897; (13) 1913
14 Map – Watson & Co. 1891

largest producer of bricks on the river, a leadership which it retained for several decades. At this moment, Ulster County was the principal center of manufacturing in the world's principal region of brickmaking, with twenty-two yards.

In 1912, manufacturers considered an agreed shutdown or severe limitation of production, with a wary eye on the recent U.S. Supreme Court decision in the Standard Oil case. *Brick And Clay Record*, also in 1912, noted the unenviable position of the Hudson River manufacturers, "who had suffered more than those in other parts of the State and in the East. . . ." Prices were under continuous pressure until 1915. The very unpleasant need for even more significant reductions in the number of manufacturers was inescapable, and this occurred soon enough. Brickmaking had become a chaotic and largely unrewarding business. It was generally failing to attract young ownership and management talent and, additionally, affecting recently hired immigrants who were forced to find other kinds of employment.

A survey of manufacturers by the New York State Department of Labor, in 1913, showed the average Haverstraw yard's employees numbering less than

Figure 49. Kiln Face. At Hutton Company, early-1900s, kiln 'setters' arranged for handing bricks up and down; bricks in the 'bottom bench' (from floor to top of arch openings) are only showing the ends with spaces between (they are not in the '3 over 3' arrangement in the bricks above that level); this arrangement allowing for maximum horizontal diffusion of heat from the firing arches, through the bottom portion of the kiln. (From DeCicco/Matthews Collection)

fifty. The Washburn Brothers of Glasco became the largest yard in number of employees (at 232) on the river, followed by the Rose Company (at 226 employees). Haverstraw had permanently lost its position of leadership.

The industry as a whole was plainly in a crisis condition. Its greatest years had just very quickly passed and its troubles were made quite public. The journal *Brick* stated, in 1910, that sixty percent of the Upriver manufacturers were shipping at a very thin profit margin that made a questionable proposition of operating at all, some of them made bricks at a loss. That same journal, in the same year, made the report on the sheer size of the Hudson River industry, which was quoted at the beginning of the Introduction to this writing, stating that the Chicago and Hudson River regions contended for the position of being the largest centers of brickmaking in the United States. The same issue of that journal (published in Chicago) went on to state that the Hudson River production was the unquestioned leader and that it was, "the largest annual producer of common brick in the world." Circumstantial evidence alone could support such a statement, insofar as the Hudson River brick industry had been furnishing the most essential material for the building of the fastest growing city in the world—at a time of enormous brick usage in the rest of the world. In addition, Ries's 1900 study of Hudson River clays stated that this region's clay, "furnishes the material for the greatest brickmaking region in either Europe or America." Those quotations can serve as the epitaph for the greatest days of the industry and described a condition of large contradictions—an enormous volume of production and perilously low prices. The year 1910 also served as a marker for the finis of decades of prodigious growth for the brick industry—averaging an increase of fifty percent for each decade over a period of fifty years. In 1910, New York State common brick prices were the lowest of any state in the Union, with the exception of Kansas. Other neighboring regions, including New England, Mid-Atlantic, and Midwestern states saw common brick prices that were ten to thirty percent higher, with the largest number at twenty percent higher.

The Greater New York was genuinely effective for a very short period of time—not more than two years. Some producers complained both of favoritism in promotion of certain brands and of assignment of advantageous barge berths to favored manufacturers. They also complained that those same favored brands were sold to the most credit-worthy dealers.

The Collapse of an Industry

The time of cooperative action by the manufacturers was very brief and they went their isolated ways, returning to the brokerages and commission men. Only seven years after the financial crisis of 1907, investor panic followed a nationwide industrial recession and a severe Wall Street bear market after the declaration of war by Austria-Hungary against Serbia. The brick industry was now fundamentally disrupted. Both 1917 and 1918 were ruinous years, only 211 million bricks being made in the latter year (sixteen percent of the zenith production year of

1906). Those two years alone witnessed the loss of twenty-two percent of the manufacturers (at sixty-six manufacturers in 1918, down from eighty-five in 1917). One of those outgoing manufacturers, the Rose Company, had been the world's largest only fifteen years prior. In 1916, a thirty percent wage increase greatly aggravated the damaging cost effects of disastrously low production and sales volume.[9] The circumstantial reason for this was as a response to the suddenly inflationary consumer price trend of the time.[10] In a period of eight years (1910-1918), the Hudson River brick industry experienced the loss of fifty percent of its manufacturers. In 1918, John B. Rose, president of the Greater New York, lost the confidence of the stockholders of that association as well as the ownership of his own business. In 1919, a Newburgh court auctioned the assets of the Rose Company brick plant at Roseton at a bankruptcy sale. By 1922, there were only fifty-two manufacturers remaining (down sixty percent from 1910). The Greater New York teetered along until it expired in 1934. It was replaced by a reborn Hudson River Brick Manufacturers Association. This became an effective service organization for representing the manufacturers in the metropolitan market for the next twenty-five years.

Salvation for Some

For those who were still in business after 1920, there was a surge of 233 percent in brick prices (from the 1907 lows) up to 1923. There was a similar increase for cement and a 36 percent increase for structural steel.[11] Such spectacular price increases must be discounted for the equally spectacular one hundred percent increase in the Consumer Price Index of 1915-1920, making the 1923 brick price (twenty dollars per thousand) closer to ten dollars in inflation-adjusted 1915 dollars.[12] Fewer manufacturers usually produce higher prices—a fact well known to the manufacturers but about which they had been previously unable to devise a workable solution.

Alternative Markets

The question must be asked regarding the Hudson River industry's adherence to the manufacture of the workhorse, jack-of-all-trades 'common' brick. Why shouldn't there have been some manufacturing of the higher priced face brick, often selling at close to triple the price of common brick? Face bricks were used for exterior, finished masonry only, because they were more particularly made and handled. C. E. Hall's 1905 survey of Hudson River brickmaking did state that the Haverstraw yard of Allison and Wood made face brick, "with a reputation beyond that of any other in the market made outside of Philadelphia. . . ." Wood was a descendant of the same James Wood, who made possible the Hudson River brick industry with his introduction of the addition of pulverized coal to the clay mixture. In the New York market of 1900, face brick utilization was significantly less than five percent of the quantity of the total brick market. The balance of the market was comprised of common brick. The big, active demand was unquestionably for common brick, and the manufacturers' understanding of that requirement was absolutely correct. The multiple of the price of face brick was a margin that any significant increase in the number of face brick manufacturers would have erased. Even nationally, the face brick market was only 4.6 percent of the total brick market. That percentage increased to 8.3 percent in 1909.

New York City also had a more extensive use of stone as a facing material. In addition, the problems caused by the calcium-contaminated clay prevented the achievement of the truly vitrified product obtained by established competitor face brick manufacturers from Perth Amboy, New Jersey, and Philadelphia (the dominant American nineteenth and early twentieth century face brickmaker)—locations without the calcium contamination of their clay. That problematical flaw in the clay may have been the most principal deterrent to most manufacturers with ambitions for entering the face brick market.

A Critic's View

It is necessary, at this point, to reply to one critic's plaint about the manufacturers' overriding concern with low cost rather than developing some unique qualitative aspect of the product. In the early-twentieth century, face bricks were no real option, and there was one other equally limited alternative that is described below. Additionally, the ability to market a substantial building product at the lowest cost in the United States (with the exception of Kansas) is definitely praiseworthy—and not pejorative, as implied by a critic of the Hudson River industry. That critic berated the manufacturers for being obsessed with low cost production—a fact of life that has already been quite sufficiently explained herein.

The Logical Persistence of an Old Technology and a Radical New Arrival

The requirement for large quantities of low cost bricks again brings up the question, already reviewed in Chapter 2, of whether any kind of kiln other than the scove type could produce the required enormous volume of bricks for an equally low price. There is no compelling evidence to indicate any possibility of meeting this particular market's demands with any alternative kiln technology available at that time. Under such conditions, the expense of installing a Hoffmann type of continuous kiln, without reasonable assurance of increased profitability, was hardly worth rational consideration. An authoritative comment on this point appeared in 1911 in the journal *Brick And Clay Record*: "In large markets where plants have enormous capacities, where fuel is high grade and relatively cheap, where the off-product can be readily sold, the continuous kiln has not made headway and cannot under the present development of that kiln. . ." This is a description of the Hudson River brick industry and its market. That same journal also cited the continuous kiln's high construction cost and its high rate of depreciation, as well as unhealthy working conditions within its chambers (during loading and unloading) that could create real problems for economical operation. The unhealthy working conditions in the continuous kiln chambers were also noted in a 1910 issue of *Brick*.

In addition, manufacturers for the booming Chicago market (second only to New York) persisted in the use of scove kilns for their large-scale production of common bricks, while also being precocious users of the latest overhead crane brick handling technology, the Penfield Setting System (labeled as revolutionary at the time). By 1912, steel framing for the support of the crane rails had actually been installed inside of existing wood kiln sheds. Only one all-steel kiln shed was under construction in that year. The radical Penfield Setting System incorporated a fork type of pickup for placing units of brick in the kiln that had been assembled in the dryers, thereby eliminating all manual setting of bricks in the kilns. This system demanded a considerable investment, in that it required substantial supplemental electrical power service. This begs the question as to why the Hudson River brickmakers had not also made use of this crane technology. A logical answer being in the chaotic state of the New York market (rapidly losing manufacturers) that understandably put an unwanted hold on such investment. In addition, all the Chicago brickmakers used the latest high-volume extrusion machinery (350,000 bricks per day per machine) that was made by Chambers Brothers of Philadelphia, one of the two mid-nineteenth-century originators of auger-type machines.

Also in extensive use were tunnel dryers. However, the technologically savvy Chicago manufacturers deliberately chose the scove kiln for the high-volume, low cost manufacture of common bricks—which were by far the largest share of their market.[13] This was in spite of the fact that there was considerable use of continuous kilns in the Midwest, a great deal of this for the manufacture of fire brick and paving brick, both of which were much higher-priced, as well as for the manufacture of face brick—for which there was a far bigger market than in New York City. The Chicago region was also fortunate to have clay with a low quartz content that permitted a rapid initial increase of kiln temperature to vitrification temperatures, thereby significantly reducing burning times and fuel consumption. This clay also had a long firing range that allowed the progressive vitrification of the clay that the calcium content denied to the Hudson River manufacturers. The foregoing was a final answer to Heinrich Ries's criticism of the persistence of scove kiln usage on the Hudson River during this era.

The Continuous Kiln in the United States

The Hoffmann type of continuous kiln was the only really viable alternative choice of kiln. Ries, perhaps carried away by the glitter of a new technology, could also be accused of insufficient understanding of the circumstance of the continuous kiln in the United States, where its European origins, requiring much smaller production quantities, created real problems in transposition to the New World, with its demand for large production capacity. A considerable number of unsatisfactory continuous kilns were built on account of those very significant difficulties in transposing European kiln design and construction practice to the United States.

A lot of those problematical kilns were built on a do-it-yourself basis, resulting in kiln wall failures from excessive movement caused by expansion and contraction of the masonry when subjected to the kilns' large temperature fluctuations that amounted to almost 2,000 degrees F. Such fluctuations required meticulous care in construction of a double-wall system that had liberal tolerance for movement of the masonry construction. Other significant problems arose in the proper design of the complex flue systems that were essential to the operation of this type of kiln. Moisture-laden air during the initial stages of firing easily obstructed such flues. In 1910, a highly successful, standardized design became available, marketed by one of the principal American clay machinery manufacturers, as described below.

Harvey Haigh's Continuous Kiln

The appearance of yet another important English contributor to American brickmaking technology—and to the history of Hudson River technological innovation—should now be noted, in the person of Harvey Haigh. This man was precocious in his rise in the business, managing a large brickmaking company at a young age in Belfast, Ireland. He was sent to the United States at the end of the nineteenth century to represent a large English brick machinery maker, using the city of Catskill as his base. Haigh proceeded to invent a recognizably improved continuous kiln, similar to the Hoffmann type. The first example of this was built in Catskill for the Kaaterskill Paving Brick Company, originally formed in 1890 as the Catskill Shale Brick Company. The use of pulverized shale for the manufacture of paving bricks, sewer pipe, etc. had a greatly increased usage during the late-nineteenth century. Paving brick manufacturing centered in the Midwest, principally in Ohio and Illinois (close to one hundred million per annum in each state around 1900). New York State production was at somewhat more than ten percent of those states' output.

The Kaaterskill Paving Brick Company made a brick from a mixture of pulverized shale and clay. Paving bricks were highly vitrified in order to minimize water absorption and to provide high strength—and they were also in a totally different market world from common bricks. In considering paving bricks as an alternative to the manufacture of common bricks, it must be kept in mind that the value of the early-twentieth century New York State paving brick market was only 1.5 percent of the common brick market value. Paving brick sold at two to three times the price of common brick, a similar premium over common brick prices as experienced with face bricks.

At that time, such use of shale in the Hudson River region was absolutely unique. Both the shale and clay (classed as Hamilton shale and Quaternary clay) were transported to the plant on a standard gauge rail line (under separate ownership) from Cairo (nine miles west of Catskill). Both of these materials were uniquely suited to the manufacture of a brick requiring a high degree of vitrification, as well as being totally atypical of what could be strictly defined as Hudson River raw materials. The railway transportation, built at no cost to that company, was absolutely critical for access to distant raw materials suitable for the manufacture of paving bricks. Rail transportation of the finished bricks was also immediately accessible, via the West Shore line. The manufacture of such bricks required a material that had a very long firing range (defined in Chapter 2)—on the order of 250 to 450 degrees F.—at five to eight times the temperature range allowed by Hudson River clays and shales.

The Kaaterskill Paving Brick Company's advertisement indicates that they also made 'building brick'—probably as a secondary product when the paving brick business was slow—and this may very well have been the higher-priced face brick. That company was the only paving brick manufacturer on the river, as well as the second largest paving brick manufacturer in New York State—and their paving bricks were reputed to be as strong as any made in the United States.

Haigh's kiln design was enormously successful. Haigh quickly collaborated in its installation, nationwide, with a principal maker of brick machinery, the American Clay Machinery Company of Bucyrus, Ohio. This firm will reappear in the following chapter of this writing as a foremost American producer of clay-working machinery. Haigh's kilns were widely installed in the United States, as well as in Mexico and South America. The Haigh kiln, in cooperation with the American Clay Machinery Company, offered a successfully standardized product that solved the design and construction problems that had plagued too many of the Hoffmann-type continuous kilns built in the United States since the late nineteenth century, thereby realizing the potential of that type of kiln. Haigh claimed Catskill as his residence for twenty-four years, eventually returning to his hometown in England.

The Kaaterskill Paving Brick Company's continuous kiln was the only example of this type that was ever built in the Hudson River region, and it certainly was known to many of the manufacturers along the river—who rejected the continuous kiln for the same reasons as did the Chicago brickmakers. The Kaaterskill Paving Brick Company also used 'beehive' (downdraft) kilns—also unique on the River. Both the continuous and beehive kilns at that plant remained singular in the region's industry. This singularity may never be adequately explained except for the unique qualities of the shale and clay used at this location, which allowed the manufacture of the highly vitrified, high-priced paving brick that was precluded to other Hudson River manufacturers because of their unavoidable reliance on typical clays of the region that had a high calcium content. As previously noted, those two alternative types of kilns could not provide the large, low cost production volumes that the dominant New York City common brick market required. As will be noted in Chapter 5, most of the economically accessible Hudson River shale also had significant calcium content, as well as other undesirable characteristics.

Tragedy in Haverstraw

Insofar as the production volume and price zenith of 1905-1906 also marked the beginning of the end for the greatest days of the industry, there was another even more dramatic and tragic harbinger in Haverstraw, the original great center of the Hudson River brick industry. As already described, the clay resources of the Bowline clay lands (between Haverstraw village and the river) had been so heavily excavated that the land surface, originally at the same height as the village, was excavated to a level lower than the river itself, a difference of fifty to sixty feet below its original surface at the Excelsior clay pits and thirty-five feet below the mean tide level at that location. As the face of the excavation moved westward, toward the village, there eventually was a conflict between the clay excavation operations and the village itself. This was duly noted in a 1902 issue of the Rockland Times, which described residents' alarm along Rockland Avenue, upon seeing a large crack in the street. Rockland Avenue, running north-south, was closest to the excavation by that date. The same issue went on to note: "Some years ago a similar crack appeared in the street and slid down into the Eckerson [owner of the Excelsior Brick Company] clay bank, taking a large part of the intersection of Rockland and Jefferson Streets. . .undoubtedly caused by the depth to which the Eckerson and Gillies excavations were made. Were the facts brought to the attention of the courts, the work of excavating would be stopped immediately."

By 1906, the Excelsior Brick Company and Gillies had undermined a considerable length of Rockland Avenue. The residents had protested the manufacturers' operations for years. One man had fought the excavations for thirty years. At this location, there is a heavy sand overburden, underlain by blue clay—an unstable condition in which the sand's weight can collapse the vertical clay excavation face which extended below the level of the river itself. The clay at the lower level was unquestionably water-saturated (pumps operating only during the brickmaking season), further destabilizing the condition.

On January 8, 1906, at 11:30 PM, an initial collapse occurred, crushing thirteen houses, followed by two more collapses. Overturned stoves started fires, and broken gas lines fed the conflagration. Nineteen people were killed and the search for bodies continued until January 27. A force of two hundred men used a derrick for recovery of bodies. All of this happened simultaneously with additional collapses of the face of the excavation. A grand jury returned an indictment against Gillies, his tenants Nicholson and Reilly, and J. Esler Eckerson, president of the Excelsior Company. The suits, totaling three hundred seventy thousand dollars (circa seven million dollars in the year 2000), were settled out of court. The brickyard owners thereupon went on land-buying binges for self-protection and to augment their clay land holdings. This tragedy also marked the final ascendancy of the Upriver manufacturers, with their seemingly inexhaustible clay resources that could be much more economically extracted. Fate marked the end of Haverstraw's dominance of the Hudson River brick industry with the tragic collapse of the excavations along Rockland Avenue. In the following year, Ulster County assumed leadership of the Hudson River brick industry.

Immigrant Workers in the Brick Industry

Haverstraw had been a source of jobs since the middle of the nineteenth century for large numbers of impoverished Irish immigrants living in the worst of the Lower East Side slums of the city. The catastrophic potato blight that began in 1846 had driven them out of Ireland, causing that country to lose almost a quarter of its population to emigration, famine, and disease. Emigration continued over a number of years as old patterns of land ownership and tenant relationships were destroyed. Early in the period of Irish immigration to the United States, considerable numbers of Irish workers went up to Haverstraw to find work in the brickyards rather than face unemployment, ruinous poverty, and unbearable living conditions in the city's worst slums. Employers often reacted harshly to workers' efforts to obtain more pay or better working conditions (including less demanding production quotas for piece work). Violence was often a result of strikes, as already described in Chapter 1. The Haverstraw brickyards were paying an 'entry level' wage to the new immigrants, and the Irish moved on to other kinds of better-paying work—or became entrepreneurs in the brick business.

Other economic and social convulsions of the European continent also drove additional waves of immigrants to the United States. The great Italian emigration was partly due to the demographic explosion that lasted from 1880 to 1915, caused far more by a rapid drop in the death rate than by a rising birth rate. The Italian population grew thirty percent in forty-five years. Most of that population pressure was in the Mezzogiorno (South). Economically damaging to the South was the post-1861 free trade policy that followed Italy's forcible unification, in effect making the South an economic colony of the North. There followed the protective tariff of 1887, which favored the industrial interests of the North, the agricultural crisis of the 1880s and a stagnant economy that was unable to absorb surplus labor. Land ownership reforms, which the great populist revolutionary leader Garibaldi promised, never materialized under the new government that the North controlled. Significant numbers of the Mezzogiorno rural population were forced to flee rising taxes that were used to benefit northern industrial growth, as well as rising prices, malaria, and poverty. Campania (including Naples), Sicily, Calabria, and Abruzzi were the provinces most affected. Large numbers of the migrants came to the United States, where industrial wages were found particularly attractive—by comparison with the miserable wages in Italy and significantly lower wages in other European countries (U.S. wages at three to four times Italian wages). The United States became "the only country which offers a sure and happy refuge for our people. . ."[14] as noted by an immigrant journal in New York City. Also noted was the fact that an American working man was entitled to consideration and respect, accords not usually provided to manual workers in Italy.

The availability of low-cost steamship passage, as well as the efficiencies of entry realized by the opening of Ellis Island in 1892, made possible the immigration of large numbers of working age males (eighty percent of all emigrants), mostly married and not from the poorest part of the population (they needed enough money to pay for the passage and to grubstake their families with enough cash to tide them over until they could send earnings back home.) Remittances from wages that were sent home, via the burgeoning immigrant banking system (Banco di Napoli, with thirty correspondents in the United States outside of New York City, the Credito Italiano, and the Banco Commerciale), became an important element in benefiting the Italian balance of payments. These men generally worked and lived in supportive enclaves composed of other Italian men, in order to accumulate enough money to allow a return to Italy with savings. Repatriation back to Italy amounted to forty-five percent from 1895 to 1900 and fifty percent during the period of 1900 to 1915, the latter group was purposefully more temporary and had a greater desire to return to Italy.

Such patterns of Italian men seeking work abroad for a number of years, with eventual return to their homes in Italy, actually persisted right up to World War II. A man could earn enough money in the United States for a return to Italy during a seasonal layoff and then return to the United States to work again—virtually commuting. Returnees to Italy had observably become accustomed to a noticeably higher standard of living. Many found their way up the river to the brickyards, with a significant number of them electing to stay in the United States, as attested by today's well-established descendants of those immigrants in the Hudson Valley.

During the post-1900 immigration period, the Hutton Company gained some of its most stalwart and productive people. At least one family eventually had three generations of men at that plant. Many came to the Ulster County yards directly from Calabria, on the basis of encouragement by earlier emigrants. Unfortunately, the arrival of many of them coincided with the beginning of very bad times in the brick business.

Hungarians were another group of immigrants who settled near the Hudson River. They were part of the five hundred thousand who fled to the United States from the harsh Hungarian 'slave law' of 1898. This law allowed labor press gangs and corporal punishment of strikers. The Jova plant was unique in that it had a workforce that was largely comprised of Hungarian immigrants. Members of this national group worked at other yards as well.

The Brickmaking Communities

Lacking the automotive mobility that allowed workers of later years to have a multiplicity of options for the location of their homes, it was necessary to live within walking distance of the brickyards during the nineteenth and early twentieth centuries. Whole communities comprised of brickyard employee families were created, as at Roseton, Brockway (one and one half miles north of Beacon), East Kingston, Glasco, and at the greatest of them all—Haverstraw. The latter community did have other manufacturing and business activity, but the brick industry dominated it. Roseton, with its three large manufacturers (Rose, Jova,

Arrow), was an isolated community. John B. Rose created a community for workers that included an attractive commons building (featuring club rooms, a reading room, and billiards) and a very proper-looking brick schoolhouse. Neither of those buildings exist today. Juan Jova's wife sponsored the 1891 construction of a notable brick Catholic church for the community (Our Lady of Mercy),[15] now isolated but still immaculately cared for (directly across River Road from Central Hudson's Roseton generating plant, which stands on the site of the Jova plant). That church today (in 2003) celebrates a well-attended Sunday mass. The majority of the congregation are of Hungarian descent, and many of those people come from some distance away.

Similar to other such communities, there was a Rose Company store that prided itself on underpricing Newburgh stores. The Jova's company store stocked imported Maillard chocolate that the Jovas purchased in quantity for consumption at the office and at home. Company stores were able to extend credit to the workers during the winter shutdown period, and other communities had stores that the brick manufacturers operated, including the Washburns at Glasco and the Brighams in East Kingston. Company stores did not have a reputation for abusive practices, unlike notorious instances in other industrial communities in United States. Company-owned housing ranged from model for the time to lodgings that were a good bit less than that. Detailed maps[16] and photographs (in C. E. Hall's *Story of Brick*) of Haverstraw brick properties show worker tenements cheek-by- jowl with brick manufacturing operations. The area of the manufacturers' property was extremely limited for almost anything other than manufacturing, causing this kind of proximity. Upriver plants had much larger properties and could locate housing at a more comfortable distance from the plant itself. The intent was to provide temporary accommodation until the families could find permanent housing in the community. The author remembers brick rental houses for workers that were one to two hundred yards away from the Hutton Company in Kingston, living conditions along North Street were still semi-rural, with city water available in 1904 and sewers in the late 1950s (as far as St. Mary's Benevolent Association).

Figure 50. Church of Our Lady of Mercy, Roseton, 1891. Construction sponsored by the wife of Juan Jacinto Jova. (Photo by author, 2001)

The Rose Company was probably the most diligent in providing a better standard of housing, insofar as that plant's location was isolated from any other community, and also because John B. Rose had an adequate share of enlightenment.

The Brockways, also in an isolated location, provided extensive housing for their workers, as did the Jovas. Occupants of the Brockway housing strongly objected to the olfactory problems that large-scale filling operations with highly questionable fill material from New York City presented. It was full of organic content and was used for extending the property out beyond the original shoreline. The composition of that fill presented disastrous problems to a successor brickmaking owner of the property—seventy years later.

Life in the Brickmaking Communities

Immigrants brought with them various needs and customs. Because so many of them were from rural backgrounds, vegetable gardening and grape culture (for homemade wine) were widely pursued. In the neighborhood near the Hutton Company, there were a half dozen outdoor brick bread ovens. One family had a resident fertilizer factory in the form of successive generations of horses, all named Cherokee. As in the case of so many other nationalities coming to this country, a benevolent association was formed for the common concerns of the new arrivals and to give them a sense of identity in a new land. Such associations were also known by the Italian name of *congreghe* and may have been inspired by the growth of such worker societies in Italy during the agricultural crisis of the 1880s. In the Italian community near the Hutton Company, the St. Mary's Benevolent Association fulfilled this function. Its initial meetings after its founding in 1911 were held in the living room of one of the founding members on North Street. The Association (originally the Benevolenza Associazione Santa Maria Di Mutuo Socorso) had thirty-two founding male members. The largest representation came from a family that had a total of six members and five other families that had two founding members from each family. The largest group was the Sottile family, a descendant of which is now mayor of the city of Kingston. A contemporary photograph of the membership shows stern-visaged men, dressed in suits, white shirts and ties, visibly reflecting their purpose of making good in a new land by dint of hard work—and they most certainly did that. The first bylaws, carefully preserved to this day, were printed in Italian. The original purposes were "to help each other in times of illness and distress," for the provision of death benefits, as well as being a center of religious and social activity—these intents being plainly stated in the society's name of "Mutuo Socorso" (mutual aid).

The association's annual September festival, complete with fireworks, has been annually observed since its founding. The society moved into its second building (built by some of the members with their own hands) in the mid-1970s—still very much alive and well in the third millennium. A striking feature of the community was the extent of mutual support and assistance in the execution of more ambitious undertakings such as extensive house repairs and construction projects—an extension of the association's principle of *mutuo socorso*. Also imported from the old country were well understood means for administering discipline to unruly children when all other reasonable means of parental persuasion had failed—such as the *schiaffo la faccia* (literally, "wash the face," delivered with the flat of the hand) and the *mano merse* (delivered with the back of the hand). The latter was used only when the grand manner was required, and it was also understood by both the disciplinarian and the recipient to cause as much pain to the former as to the latter.

A memorable custom that was common to many of the brickmaking communities was 'carrying the pail.' This was the function of children in the nearby grammar school, hurrying home while on their lunch hour, who would then take their fathers' freshly prepared lunch to them, arriving at the plants just before the noon whistle. Some children carried several lunch pails—for other men whose children had gone beyond grammar school—and the watch could be set by the file of children moving down through the plant with the lunch pails. The 'pail' was named after the original shape of the container, which was a vertical metal cylinder with a bail handle. The pasta was placed in the bottom, followed by other contents, with the dessert usually on top. The horizontal lunch bucket came later, accompanied by the mythology of a man being able to carry two bricks home in the lunch bucket every day, resulting in fine brick utility sheds, brick veneered house walls, etc.

The custom of 'carrying the pail' may well have originated during the old brickyard days with their crack-of-dawn starting times. Some children did this on a business basis. Vincent DeLuca, during his East Kingston boyhood, carried Ed Post's lunch pail for one dollar per week, and Ed always saved some dessert for the young man. Vincent, today a Kingston businessman, also gives his own account of the brickyard community of East Kingston. His account is notable for his frank affection and admiration for a tightly-knit village of human beings living solid, hard-working lives together during the pre-World War II years. (In 1913, there were 354 brickyard workers in five yards at East Kingston.) Vincent's grandfather, Vincenzo Pietro DeLuca, from near Cosenza (Calabria), emigrated to the United States in 1904, initially working on railroads near Chicago as a dynamiter. Vincenzo married another transplanted Calabrese, Josephine Castiglione.

East Kingston had the benefit of two benevolent societies. The earliest was the Society of St. Liberata. That saint was venerated around a particular locality in Calabria. Somewhat later, the Society of St. John was founded. The two societies were amalgamated in the 1950s and disbanded in 1999. Similar to other such benevolent societies, funds were collected for men who were out of work on account of illness or injury, as well as death benefits. Those societies also had annual festivals featuring elaborate fireworks that were prepared with the assistance of members and their older children. They fastened rockets onto sticks that were then set in a trench, so as to be fired off simultaneously, revealing an American flag. Trenches also were dug for spectacular mortars. The festivals were scheduled to avoid conflict, with the St. John society's in early June, followed by St. Liberata in early September. They were always observed on a Saturday, and the fireworks began at the stroke of midnight. There was so much fireworks business with those communities that it created local fireworks specialists like Louis Cosenza of Ulster Landing.

Following mass on the following Sunday morning, the men of the society bid for the honor of being one of the four bearers of the statue of their society's saint. The statue was carried through the village, in the company of a band, with a stop in front of each house. The lady of the house then emerged to place a ribbon bearing a donation on the statue. The boys fiercely competed for the job of carrying the front end of the band's big bass drum. The love of fireworks extended to the borrowed pleasure and anticipation of watching the great annual Fourth of July display that was put up on the Astor estate. It was plainly visible to the south on the opposite shore of the river, and the spectacle lasted for over an hour. During the winter, there was a great need for ice harvest workers from the village. The men bought rubber boots with heavy felt insoles, purchasing them two sizes bigger in order to stuff them with newspaper for insulation. It could be very cold and windy out on the river—and the horses could fall through the ice, requiring special techniques for rescue. In addition to the Brigham Company's store, which extended credit during the winter layoff time, local residents owned several small stores specializing in meat and sausage, etc. Gardens were *de rigeur*. Two hundred fifty to three hundred tomato plants in a garden was usual, as was home canning of garden produce. The forgoing is a glimpse of life in East Kingston—a village remarkable for its cohesiveness, mutual cooperation and respect among its members, which included Germans, Italians, Polish, and African Americans.

Glasco was dominated by the Washburn Brothers' two large brickyards and its Italian immigrant population, which was preceded by Germans. The presence of brickmaking in Glasco was inescapable, with the Lent yard's operations straddling the village's main thoroughfare to the river (Delaware Street). The Washburns' big company store was a true department store and was notable for the fairness of its dealings with the community. That same sense of concern was extended to other forms of assistance to the citizens of Glasco. In 1867 (date of founding of their Glasco brick business), the Washburn brothers built a five-room schoolhouse for the community and later sponsored a park, which featured weekly band concerts. The Washburns also owned forty to fifty properties in Glasco, and as late as the 1930s and 1940s, company housing was rented for two dollars per month, with paint and essential maintenance furnished gratis. Members of the community were largely of Calabrian origins, with the exception of one group of four families from the province of Molise (southeast of Rome). One of those families, the Ciarlantes, quickly caught the entrepreneurial spirit of their new land and developed a substantial coal and ice business. Antonio Ciarlante, an early arrival in 1880s, decided that the United States was the future's best hope.

He thereupon returned to Italy to marry, and then permanently returned to the United States.

Joseph Mayone's life was singular among the new arrivals in that he got off to a fast start. Born in 1865, he worked as a water carrier in Algiers for the French government during his teen years, coming to the United States in the mid-1880s, not yet twenty years of age. He proceeded to lease a brickyard in Portersville (just south of Glasco), subsequently building a yard in Athens ca. 1910, and later in Catskill. With ownership of a store in Glasco, Mayone extended credit to local workers, many of whom paid up their accounts by working on his new house in Saugerties (complete with gardens and fountains). He also became a director of the bank in Athens. In all, he had a record of making the very best of the opportunities the New World afforded.

Glasco experienced the same spirit of mutual assistance among its families as in other similar communities, with an institutional presence in the form of St. Joseph's Church—in lieu of the benevolent societies of Kingston and East Kingston. The Washburn Brothers' brick plants observed as a holiday Glasco's annual festival, the Feast of the Assumption, celebrated on August 15. Skills for day-to-day living, brought over from the old country, included the baking of bread in outdoor ovens, extensive gardening, and significant hog husbandry. Slaughtering commenced upon the arrival of cold weather. Imported skills were utilized for meat preservation in the form of sausage making, ham curing, etc. With the nearest banks in Saugerties, at a time when few of the workers owned automobiles, many of the workers entrusted their money to one of the stalwart women in the community who became an in-town savings institution. There was also a significant practice of midwifery in Glasco. In all, those communities were a tribute to the ability of new arrivals to this country to create a highly successful communal existence in the New World.

Paralyzed Pigs

Vincent DeLuca also tells a splendid tale of adaptive reuse of a brickyard feature, namely the brick flue of the abandoned East Kingston Terry plant. Some local entrepreneurs in the business of manufacturing beverages that busybody laws controlled during Prohibition enlisted this structure, approximately sixty to seventy feet in height. The flue served to remove fumes from the distillation process to an altitude from which they could be dispersed without human detection. With the sour mash byproduct still requiring disposal, a neighboring pig farmer was prevailed upon to allow dumping of the mash on the edge of his property. Upon seeing his pigs lying helplessly paralyzed on the ground and fearing for their very lives, the farmer called in a veterinarian who diagnosed the pigs as being stupefied by the alcohol in the fermented mash. The veterinarian immediately notified relevant law officers. The area was staked out by said law officers to await the next mash deposit. The beverage producers had a prosperous business enterprise curtailed by unsympathetic laws.

Seasonal Workers from the American South

Another group of workers that began to figure prominently in the industry at the turn of the nineteenth century were African American. A large number of these men, initially from rural Virginia backgrounds, were hired on a seasonal basis in the Haverstraw yards, beginning around 1900. A brick company had a contact in the Virginia community who obtained assents from the individual worker prospects, whereupon the men gathered at a designated location for tickets to be issued for their transportation north. Such recruitment could be at odds with laws of certain states that forbade such interstate labor transportation,[17] and men from plantations had to get the plantation owners' permission to leave for the season. By one account, the state of Virginia required a five-thousand-dollar license fee for the recruitment of workers.

The employment period extended into the early fall. The workers lived together in large camps. At some point in time, this practice spread to the Upriver yards. Some of those workers chose to become permanent residents in the Hudson Valley, with concentrations of descendants today living in Beacon. Workers from certain southern states tended to work in the same brickmaking communities. Haverstraw had the biggest representation from various states, including North Carolina, Georgia, Virginia, Maryland, and eastern Alabama. Virginians went to Coxsackie, Coeymans and Kingston. Men from North Carolina went to Hudson, Coeymans, and Beacon.

Hiring agents from those southern communities, often educated men, initially hired relatives. It must be understood that this was considered desirable work. Women looking for a husbands regarded brickyard workers in those communities as a good catch, and the same went for railroad workers. Many of those who had permanently moved into the northern communities provided accommodation and services for the seasonal workers. Permanent residents started churches, restaurants, and bars for people in that community. In the case of one Upriver community, African American housing was isolated on an island in the river, with a store in the same location—a bit of Jim Crow in the north.

The use of such seasonal labor ceased during the 1920s and 1930s after replacement of the old wood kiln buildings with steel sheds that had overhead cranes. This allowed for new labor-saving brick handling methods. At the Hutton Company, other very productive men—who happened to be African American—became part of the permanent workforce. At least one of them had a job with considerable responsibility, this in the 1950s.

One Plant's Families

At the Hutton Company, the arrival of a number of men, whose presence was of real value to the company, marked the first decade of the twentieth century. This presence lasted right into the middle of the century. One of these men was Anthony Polacco, another arrival from near Cosenza in 1909, who was the earliest of three generations of Polaccos, who were with the company for the next sixty years. Others included Frank Turck and Sam Frangello (arriving in 1904) whose sons were also with the company, in addition to the first of the Naccaratos, Perrys, Sottiles, Ferraros, and many others. Frank Turck's son, Thomas, became both superintendent and a director of the Hutton Company. In his search for a better life, Anthony Polacco first went to South America, returning to Italy before setting forth for the United States. He sent for his family after two years of residence in his adopted country.

Two more men to have important roles in the plant's operations were William Tierney, arriving in 1906, who became plant superintendent—and the first non-family director of the company—and Jack Sanglyn, arriving in 1909, who shortly thereafter became a kiln shed foreman. Both of those gentlemen represented an echo of the old Irish foreman days in the industry, which included such men as Maurice Duffy, another first rate who also became a kiln shed foreman. Tierney and Sanglyn were with the company into the middle of the century, with Maurice Duffy present until the plant's closing in 1980.

Effects of Wold War I

At the peak of U.S. involvement in World War I, the slowing of the brick industry saw the Hutton Company leasing its barges in 1918. The desperate need for 'lighters' (vessels for transportation of cargoes between docks and moored ships) occasioned this situation. Enormous shipping activity during World War I brought this on. The brick business was a disaster at that moment—so barge leasing may have been the only profitable niche available. Directors' minutes of the following year show the development of good interbusiness relations with the presentation of gifts of cigars to Cornell tugboat captains—an idea sure to horrify present-day sensibilities, with all the implications of unintentional inhalation of tobacco smoke in a confined space (the tug's small cabin) by nonsmokers (if there were any), etc. Directors' minutes of 1919 also showed the purchase of additional electric motors for brick machine power, demonstrating an improvisatory approach to the upgrading of facilities, rather than the kind of wholesale modernization that completely remade the plant during the next decade. A note in the Directors' minutes of 1906 also mentioned a cash refund for the unused ration of oats and hay for the horse of one of the directors—such horse feed was a board perk of the time.

A Farewell to Old Ways

At its early-twentieth-century zenith, the Hudson River brick industry employed between seven and eight thousand workers.[18] Before mechanization in the 1920s, brickyard work was generally regarded as hard, manual labor subject to seasonal shutdown, using both strong mens' backs, steam power, and horsepower to get the work done. Tragically, the economic predicament of the industry, during the early-twentieth century, created a condition of low wages that gave brickyard work an unenviable reputation. That condition significantly improved after the coming mechanization, along with a significant reduction in the number of manufacturers, the unionization of the 1930s and substantial post-World War II wage increases. Note should be made of an authoritative journal's report that the brick industry's rate of injuries, in the early twentieth century, was below the injury rate in other industries.

The Hutton Company was now to enter into a very different existence from anything it had experienced. Before the 1887 cessation of the Cordts and Hutton partnership, due to Cordts's retirement, John H. Cordts's name was most prominently attached to the yard, with the yard's location given as the business address. Cordts died in 1891. The Cordts family continued to live in the mansion that Cordts built. His son, John N. Cordts, not only pursued a very successful business career but also became a U.S. presidential elector for Theodore Roosevelt, as well as a thrice-elected New York State senator.

After 1890, the brickyard's business address was at William Hutton's East Strand lumber yard office in Rondout, over a mile from the brick yard—and it remained there until 1921. Such a remote location for an office from the brickyard wasn't unique, in that the Staples yard's office was just up the hill from Rondout. Their yard was almost two miles away at the north end of the Kingston shoreline (adjacent to East Kingston).

The period of time covered by this chapter, 1897 to 1920, was one of moderate leadership and moderate investment at the Hutton Company. The company did not conspicuously bear the stamp of the owner, as did other companies such as Jova, Washburn, Shultz, Terry, and Rose. However, the company did manage to not only stay in business during trying times but also to retain its position as one of the principal manufacturers on the river. The termination of this chapter's time period is a coincidence of two events, thereby marking the beginning of an entirely new era for both the Hudson River brick business and also for the Hutton Company. The cause of the latter was the death of George Hutton, secretary and treasurer of the company, in December 1920. The result of this event was the immediate entry of his son into the business, the consequence of which was radical change. The pivotal circumstance in the industry's fate was the beginning of the 1920s. At the end of this period, the stock market crash that brought on the Great Depression both transformed and traumatized the Hudson River brick industry along with the rest of the nation.

Notes

1. *Clay Industries of New York*. Heinrich Ries, 1895.
2. *Catskill Daily Mail*. 1987.
3. *Clays of New York*. Heinrich Ries, 1900.
4. *The Story of Brick*. C. E. Hall.
5. See Fig.48. Ulster County Brickyards.

6. Matter of William Hutton. Audit of Estate. Ulster County Surrogate Court.
7. *Brick*. Oct. 1910.
8. New York State Dept. of Labor. 1913.
9. *Brick and Clay Record*. May 1916.
10. CPI increase of ten percent in one year 1915-1916 (U.S. Dept. of Labor Consumer Price Index).
11. Prices for common brick, cement, and structural steel were annual prices at New York, furnished by *Engineering News Record*, commencing in 1874; brick prices were wholesale f.o.b.
12. Per U.S. Dept. of Labor CPI .
13. *Brick and Clay Record*. Feb. 1912.
14. In *Il Proletario* (Italian Socialist newspaper)—published in *Charities XII*, May 1904, New York.
15. Figure 50. Photograph 2001 (Our Lady of Mercy).
16. Sanborn-Perris Insurance Maps (Figure 35).
17. deNoyelles.
18. On the basis of an average of sixty workers per yard.
19. *Clay Record*. Jan. 1908.
20. A recent writing stated that there was a "pervasive" presence of airborne dust in both the brick and cement industries and that respiratory ailments were a common occurrences among workers, whose sole protection was the use of wet handkerchiefs across the face. Contradicting that statement is testimony by one member (himself with the brick business for thirty-five years) of a family representing three generations of workers in the brick industry. He confirmed that airborne dust occurred infrequently and only in the immediate vicinity of mold-dusting operations. The author's own experience at the plant supports this statement. Cement workers engaged in quarrying operations did work in caves where restricted conditions may have created adverse conditions with airborne dust (from blasting, etc.). In the old yards, brick molding machines were provided with roof coverage only (usually minimal in area)—hardly restricted atmospheric conditions, and there could not possibly have been any accumulation of airborne dust in the open yards or in the kiln sheds—because there was no source of such dust in those locations. In addition, William Minnock (see Bibliography) states that, during the late 1940s, the New York State Department of Labor checked machine room personnel for pulmonary disorders and issued a negative finding after an inquiry. Due to lack of evidence, the investigation was terminated. The only recommended precautionary measure was for the worker responsible for mold dusting to wear a respirator while performing that task.

Principal Sources for Chapter 3

American Clay Machinery Co. *Catalog No. 90.*
American Institute of Architects. *Guide to New York City.*
Brick. March 1905, May 1908, Oct. 1910.
Brick and Clay Record. June, July 1911; Jan., Feb., July 1912; April, May 1916; July 1918.
Clay Record. Jan. 1908.
Clay-Worker. Jan.-June 1904 .

Hall, C. E. *The Story of Brick.*
Engineering News Record. *Construction Costs.*
Hutton Co. Minutes of the Stockholders and Corporate Directors.
Museum Bulletins of New York State. Mining and Quarry series. #47, #120, #166, #178.
Ries, Heinrich. *Clays of the United States.*
_____. *Clays of New York.*
Rockland Messenger. (See Bibliography).
Rosenwaike, Ira. *Population History of New York City.*
Swain, Charles.
U.S. Dept. of Labor. *Consumer Price Index.*
U.S. Patent Office. (See Appendix).
U.S. Bureau of Census. *Historical Statistics* (Cost of Living Index).
WPA. *Guide to New York City.*

CHAPTER 4
Becoming a Modern Industry:
The Quick and the Dead—1920–1941

FROM THE BEGINNING OF 1921 UNTIL 1965, the fortunes of the Hutton Company were tied to the new member of the firm who joined the company upon his father's death in December 1920. This person was George Van Deusen Hutton, hereafter referred to as GVDH or Van Hutton. Born in 1897, his father was George Hutton, treasurer and secretary of the Hutton Company, and his mother Augusta Van Deusen, a daughter of Kingston's principal druggist family, makers of *Van Deusen's Ready Remedy, Russian Horse Liniment*, and other patent medicines. The latter was a very prosperous business. An uncle of Augusta's built the showy 'Cloverly' in the mid-1890s, farther up on 'the Weinberg' (West Chestnut Street) from William Hutton's house.

At age nine, Van Hutton contracted a severe case of poliomyelitis, the consequences of which burdened him for the rest of his life and created within him a case-hardened willpower and an intellect to match (honor student at Phillips

Figure 51. George Van Deusen Hutton, 1920. (Williams College graduation photograph)

Exeter Academy and Phi Beta Kappa at Williams College, class of 1920). Personal misfortune equipped him with a keen sensitivity to the condition of others who had experienced the arbitrarily harsh workings of fate and an admiration for those who fought to overcome them. Quite opposite to having any intentions of going into the brick business, he could see its troubled ways and backwardness. Rather, his intentions of going into the New York City banking business (a fast track of the time) was plainly stated in his college yearbook.[1] Six months after graduating from college, immediately following his father's death from a fatal illness, family members pleaded with him to enter the brick business. In order to obtain his acquiescence, there had to be a *quid pro quo*—joining the firm implicitly required that there had to be radical change in the manner in which the company made bricks and conducted its business—and adequate funds from earnings had to be furnished for that purpose. The sense of duty prevailed over previous feelings about the brickyard—and any other ambitions—and he henceforth gave the firm, unremittingly, his considerable talents and tenacity until his death in 1965. In the same month as his father's death, he was made a director and secretary, becoming treasurer and general manager in the following year. He retained the treasurer position until his death in 1965.

New Talents at Work

Even though lacking the title of president for another five years, Van Hutton acted *de facto* in that capacity and lost no time in moving the company's office to the manufacturing property, into a small new (brick) office building. He also brought a new principal figure into the life of the Hutton Company, with the arrival in 1921 of Montgomery Burhans—who many times over confirmed GVDH's opinion of him as a mechanical genius.

The first Burhans in Ulster County was Jacobus, who appeared in Esopus in 1660 as a soldier in the service of Holland's director general, and who had the bad luck of losing two houses in the 1663 burning of Kingston (named Wildwyck at that time) by the Indians. 'Mont' Burhans applied his mechanical talents at the Hutton Company for almost forty years. GVDH held him in high regard, and Burhans later became a director of the company.

In the way of bumps and grinds of a family business, one of the family directors persistently withheld votes on important matters relating to the new management, giving no explanation. One of GVDH's earliest acts was to arrange for the sale of company property to the St. Mary's Benevolent Society for a permanent building, the price to be negotiated by himself, as noted in directors' minutes. In such an instance, that could only mean that the price was very reasonable indeed (if there was a price) and he wouldn't need to argue about it with any other directors.

The New Order

Fortune was extraordinarily kind to the ambitions of the new firm member, after getting past the short but severe depression of 1920-1921 (caused by World War I hyperinflation, a newly established Federal Reserve Bank's drastic interest rate increase, as well as twelve percent unemployment) that caused a short-term drop in brick prices and production. After that, the New York City metropolitan area went on one of the largest construction binges in its history, driving up the price of brick to twenty dollars per thousand in 1923.[2] That price was a short-lived spike, and an up and down descent to twelve dollars in 1928 followed. The Consumer Price Index was flat during this period of time.

The great building boom faded even before the Great Depression, with a twenty-five percent drop in the number of building permits issued in the three years following 1925. The large scale closure of over half of the yards on the river, during the previous decade, as well as big population relocations (requiring much new construction), in Queens (up 130 percent), in The Bronx (up 72 percent) and more moderately in Brooklyn (up 26 percent), plus an enormous amount of commercial construction aided profitability during the 1920s. The dispersing effects of the new subway and commuter rail systems were also exerting a new influence on population location that rapidly accelerated, and this was augment by the arrival of widespread automobile ownership.

The old center of population growth—and principal brick market, Manhattan, had actually experienced an outright reduction of eighteen percent in population during the 1920s. The Hutton Company supplied bricks to distinguished projects in this period, such as Holland House in Forest Hills (Queens),[3] notable for its handsomely detailed brickwork.

Much Better Times

The only Hutton Company records available for the period showed a twenty-percent increase in gross revenues in the years 1925-1926. Half of this increase came from a ten-percent market price increase. The effect of prices that allowed genuinely profitable operation created enough capital for the progressive Hudson River brick manufacturers to make the investments in plant modernization that allowed survival through the Great Depression. Such efficiencies also absorbed increased costs due to unionization in the 1930s. This was time of the quick and the dead.

Figure 52. Holland House, Forest Hills (Queens). Built in 1920s with Hutton Company bricks; note finely detailed brickwork, with shadow course panels below windows and closely spaced projecting headers adding shadow interest to wall surfaces. (Photo by F. C. Monteino)

Like several other plants, the Hutton Company seized the opportunity to aggressively turn its back on the traditional manufacturing methods and build new, twentieth century facilities that henceforth changed the name of the place of manufacturing from brick*yard* to brick*plant*. Substantial investment did not necessarily guarantee survival, as will be seen, but survival was impossible without it, insofar as the new investments significantly increased productivity. Concomitant with the increase in productivity was a reduced payroll—fewer men employed. Various signs of changeover included a 1926 trade journal advertisement selling Dennings Point's newly castoff three-hundred-horsepower Corliss steam engine (undoubtedly from the next-door factory of the Fishkill Landing Machine Company). This resulted from the complete the modernization of that plant in the same year, as described below. Notable in the process of investment in modernization of principal manufacturing equipment is the cessation of the purchase of such equipment from Hudson River machinemakers, which in the nineteenth century manufactured not only brick machines but steam engines as well. Machinery for modernization of the Hudson River plants henceforth came from out-of-state, with limited exceptions.

Testing the New Talent

There remained the question of how a very capable, but young and inexperienced new manager would decide on the needed investments, with little in the way of genuinely effective in-house assistance. Van Hutton was essentially on his own in a tough industry, at age twenty-three. A very quick study of available possibilities was required—and the decision to aggressively prosecute an investment program, with no delay, was the correct one. Recalling the problems presented by the local clay quality, the choice of technologies was restricted with respect to forming the bricks, in being limited to the soft mud molding process. The drying process sorely needed change, and there were at least two drying technologies available. The only other major possibility of change was in the type of kiln—abandoning the scove kiln for a continuous type. In this latter category, there existed a very promising new possibility in the tunnel kiln—a new continuous kiln technology that very quickly replaced the Hoffmann type. However, tunnel kilns only appeared at one Hudson River plant during this period of time. The other plants that pursued aggressive investment made very similar choices in the selection of new equipment.

The Investment Program

The investment process at the Hutton Company began in earnest in 1922, with the start of a three-year program of replacement of the open yard drying process with steam drying. Eliminating that quaint carryover from ages past was at the top of every Hudson River brickmaker's list of improvements. The drying system[4] included four new buildings with steam pipe dryer racks, plus a cable conveyor system for transporting steel pallets and a separate boiler building for provision of steam. Pallets were manually removed from the conveyor and shoved into the spaces between the horizontal steam pipes. This dryer system became the standard for the Hudson River plants, partly because of the manner in which the principal molded brick machinery makers provided comprehensive plant building services that included steam drying. Lancaster Iron Works (Lancaster, Pennsylvania) was responsible for doing the entire new installation of conveyors from brick machine to dryers, as well as the design of the dryer 'houses.' This firm figured prominently in the modernization of not only the Hutton Company but of

Figure 53. Steam Dryer Designed by Lancaster Iron Works, 1920s. The cable conveyor system and steam pipe racks support steel pallets and bricks. (From advertisement in Brick and Clay Record*)*

the majority of the other modernizing plants on the river as well. Lancaster Iron Works claimed the first American installation of such pipe rack dryers in 1898, as well as the first cable conveyor system in 1900. The steam dryer system allowed the drying of bricks in twelve to sixteen hours instead of the three days required for open-yard drying.

Dealing With the Seemingly Impossible

The new management also identified a very immediate source of crippling weakness that would be fatal to the company. The company was about to run out of clay. A detailed topographical survey of the clay banks, dated 1923, which GVDH unquestionably ordered, established this beyond any doubt. It showed the extent of excavations relative to the west property line (the old Moses Cantine Patent line). The top edge of the excavation was right against the patent line for most of its length, with a slight dip into the adjacent property at one point. Recall that Sylvester's 1880 *History of Ulster County* had caroled about "an inexhaustible bank of clay." Exhaustion was fast approaching, but fortunately a remedy was available. That remedy lay in the immediate (1923) purchase of the abutting property, to the west of the Moses Cantine Patent line, from the corporate remains of the Newark Lime and Cement Company, once a big presence in Ponckhockie and Rondout. Commencing operations in Kingston in 1851, the local operation of that company produced what was known as 'natural cement' until 1905. Its demise was attributable to the rapidly increasing use of the much faster setting Portland cement.

The children's Sunday School Church of 1870 (now Congregational), built in neighboring Ponckhockie by the Newark Lime and Cement Company, was one of the very earliest nonresidential concrete buildings in the United States. It is an imposing monument of pioneering concrete construction to this day. The architectural historian-critic, Ada Louise Huxtable, cited that building in a 1960 article on notable early American concrete structures in *Progressive Architecture* magazine. In the same community, also of concrete and built by the Newark Lime and Cement Company, is the former Union Free School of 1867, still standing. Many generations of Cordts and Hutton, as well as Hutton Company children attended classes in that building for almost one hundred years.

The Hutton Company purchased fifteen acres of badly needed clay land. The Terry plant immediately to the north—also critically low on clay resources—previously acquired additional clay property from the same source. Within nine years, the new clay land operations presented a threat to the existence of the Hutton Company.

Wholesale Mechanization

The widespread use of electricity in the 1920s had a powerful effect on productivity, strongly affecting the Hudson River brick industry including the Hutton

Company. The years 1923 to 1925 saw the purchase of four new THEW (that name stenciled in very large letters on the cab) electric-powered clay-and-sand excavating shovels (manufactured by the Thew Automatic Shovel Company of Lorain, Ohio), thus ending the hard manual labor of heaving the clay into horse-drawn clay carts. Previous to manufacturing electric shovels, Thew had been a major manufacturer of steam-powered excavating equipment. The year 1924 also saw the abandonment of coal-fired kilns for oil firing. A not-too-fond farewell to the horse-drawn clay carts came the following year. The rapid expansion of the use of internal combustion engines in the United States, during the 1920s, also had a local presence. The plant installed a narrow-gauge rail line to the clay banks, with two gasoline-powered locomotives (manufactured by Vulcan Iron Works of Wilkes-Barre, Pennsylvania) and side-dump cars imported from England (Figure 54). With their own enormous brick-manufacturing industry, the English were principal innovators and exporters of brickmaking equipment. A full-page advertisement by Vulcan Iron Works featured a letter (dated 1926), from GVDH, testifying to the excellence of the locomotives in successive issues of the journal *Brick and Clay Record*. That letter also stated that the plant was operating at the rate of one hundred sixty thousand bricks per day—an amount that could only have been obtained by some arrangement of double-shifting. Times were good and 1926 was the last time that the Hudson River brick industry attained the billion bricks mark for one year's production.

Figure 54. Locomotives and Clay Cars at Hutton Company, 1926. Dryer 'houses' (with roof monitors) at left and new building for brick machine at right. (Photo by Vulcan Iron Works)

An Echo of a Departed Industry

After five years of Van Hutton functioning as president of the company without the title, conferrence of that position came to pass in 1926. In the previous year, he also got married–to Ruth Whitney Shafer. The spelling of the name was changed at some time in the nineteenth century from the Dutch spelling of Shaver (the letter 'v' is pronounced in Dutch as an 'f', as it also is in German). Shavertown was the name of the ancestral village, founded in the 1780s, near the village of Andes in the Delaware County. The construction of the Pepacton Reservoir in the 1950s obliterated Shavertown. Ruth Whitney Shafer's father and grandfather were in the cement business in Rosendale (the partnership of Connely and Shafer). They were one of a number of successful bidders to supply cement for the construction of the Brooklyn Bridge, as well as provisioners to boatmen on the Delaware and Hudson Canal. (An R. G. Dun and Company credit report of 1878 gave Connely and Shafer high marks: "first class businessmen—reliable and honest and will pay as long as they have a cent left.") Ruth also served as a director of the Hutton Company.

Other Modernization Programs

More preferable than the incremental additions of the Hutton Company might have been a new plant, fully designed and executed at one time, as at the Staples' Malden plant which the noted inventor, David Strickland designed and the Wellington Machine Company built in 1920 (Figure 55). That plant's modernization stopped short of building a steel kiln shed with overhead crane and related gantry, but retained its traditional wood shed with the addition of an abbreviated gantry (fixed in one location at dock) for picking up brick from a truck for loading on the boat (Figure 56). The expense of buying all the other equipment and buildings at once probably occasioned this kind of shortfall. Van Hutton's pay-as-you-go mentality did increase the chances for creating coordination problems, and there was one minor error in matching later equipment installation with earlier equipment, which eventually was problematic to some limited degree. Somehow, it all fitted together with considerable success. It was an ongoing process of persistent and consistent development over a period of seventeen years.

Engineering drawings of 1928 (at the Hutton Company) for "A Brick-Handling Gantry," similar to the installation at the Staples' Malden plant (see above), was evidence of another brickhandling option. That option was rejected in favor of a 1928 purchase of a rail-mounted hammerhead crane,[5] manufactured by the Lidgerwood Manufacturing Company (Elizabeth, New Jersey), a company that was a manufacturer of notably ingenious equipment for the construction of the Panama Canal (the Lidgerwood System). This innovation replaced three hundred men in the work of rapidly clearing rail cars of excavation rubble taken from the canal cut.[6] The name LIDGERWOOD was made very visible on a large steel panel, projecting above the top of the control cab, bearing the manufacturer's name in bold, cutout letters. (The author grew up in the presence of all those ringing names of earlier twentieth-century industrial technology—they are hard to

Figure 55. Staples Brick Plant at Malden, early 1920s. Designed by David Strickland; new brick machine building at left, new dryer building at center and old kiln shed at right. (Photo in collection of Hudson River Maritime Museum; copy by John F. Matthews)

Figure 56, below. Staples Brick Plant at Malden. Oold kiln shed (with portion of roof boards removed, at left, for kiln burning), this kiln shed remained operational until the plant closed in 1948; the free-standing gantry crane (in front of shed) is that plant's solution to barge loading, and was still in place in 2000. (Photo in collection of Hudson River Maritime Museum; copy by John F. Matthews)

forget!) The rails allowed this crane to serve the entire length of the old wood kiln shed. That crane stayed in service for five decades. (In the year 2003, it still stands.) The Lidgerwood crane was a somewhat puzzling purchase, insofar as there was a directors' authorization in 1926 for the president "to proceed at his discretion with the erection at a cost of approximately $100,000 of steel kiln sheds and unit handling system." That statement also indicated GVDH's watchfulness for technological improvements that other manufacturers installed—in this case the Jova plant with its full-blown new shed and crane system, as well as Dennings Point's wholesale rebuilding of 1926. This was not only a far more successful solution to kiln building and barge loading, but also far more expensive.

Figure 57. Lidgerwood Crane Installed in 1928. (Photo by Hutton Company)

The Lidgerwood crane did at least put an end to the arduous task of loading barges by wheelbarrow. There were two reasons for delaying the steel shed purchase—one was the priority requirement for a modern, high volume brick machine—and the other reason was what can only be described as a characterological strength of the new president. This can best be summarized by a line in his tenth college reunion souvenir book of 1930 (only seven months after the Great Stock Market Crash of 1929), stating that "Van Hutton is probably the only man in the Class who didn't buy stock on margin." This entry sounds somewhat rueful and perhaps a bit envious. The class agent must have taken a poll, or else GVDH had a reputation. In other words, cash is best, and there was enough cash generated to accomplish a great deal. He wasn't alone in this regard. Contrary to today's preference for big debt, funding of capital expenses from retained earnings was also the preferred investment strategy of such illustrious industrialists as Andrew Carnegie and John D. Rockefeller.[7]

In a nice switch from the previous decade, the 1920s were great for the profitability that allowed the investments necessary for survival. Judging from the limited amount of financial information available, GVDH liked sizeable, liquid reserves and being debt-free. This instinct was unquestionably of great assistance in allowing the company to survive the Great Depression, which broke many other manufacturers, including some of the biggest such as Brockway and Excelsior.

New Brickmaking Machinery

The final Hutton Company purchase of the 1920s, was the 1928 installation of the very latest high-volume, automatic molded brick machine, purchased from the principal supplier to the modernizing Hudson River plants, the Lancaster Iron Works. Hudson River plants that purchased Lancaster Iron Works's machines included Empire (Columbia County), Powell and Minnock, and Jova. The latter retained a somewhat older machine that was affectionately named 'Little Lady.'

The multitude of old brick machines had to go, with all their multiplicity of increasingly expensive manpower and problems of control of mixes—and there were approximately a dozen such machines at the Hutton Company. The installation included a new building that had all necessary conveyors for introducing materials into the brick machine and for removing the newly molded bricks to the dryers, with powerful electric motors for driving the various parts of this system. One of the principal benefits in the use of a single brick machine in lieu of the larger number of older machines lay in the centralization of controlling the introduction of the constituent materials into the mixture, as well as the great increase in efficiency due to reduction of manpower. Coincidental with the new brick machine's installation was the very astute installation of two granulators (trenches with rotating knives for initially breaking up the material from the clay cars), symmetrically arranged at the end of the conveyor to the brick machine, with the granulators feeding toward that conveyor (Figure 58).[8] Such an arrangement allowed the controlled introduction of clay of differing quality from various parts of the clay bank—and there was significant variation. Those dual granulators may have been unique in the Hudson River industry and they became very critical indeed during the 1950s. Mixture control was critical to brick quality. Various means were used to accomplish this, with various results. The necessary addition of sand increased the need for controls of this additive. Some manufacturers added sand directly to the clay cars before adding the clay—an insufficiently precise procedure. The Hutton Company used separate cars for clay and sand. Damp sand was dumped into a heated hopper for moisture removal, allowing better control of moisture content. A conveyor then fed sand to hoppers that were positioned over the granulators, where its addition was carefully controlled.[9]

Another Distinguished Hudson River Inventor

David Strickland invented one of the very successful earlier automatic molded brick machines of that time. His highly successful steam dryer system design had been introduced in 1912. Strickland was a prolific inventor, with sixteen patents in his name during his lifetime. A farmer's son, born in Haverstraw in 1883, he learned the brick business by working at many of the jobs in the Haverstraw yards, including machine boy, brick setter, brick burner, foreman and stationary engineer (steam engine operator). Strickland acquired his brick machinery design experience through an apprenticeship in 1910 at the Wiles foundry at Grassy

Point (cited in Chapter 1 as a principal brick machinery maker).

Within two years, his highly successful steam dryer design of 1912 appeared in a New Jersey yard, followed by his association with the largest clay working machinery manufacturer in the United States, the American Clay Machinery Company of Bucyrus, Ohio (cited in the previous chapter for its collaboration in manufacturing Haigh continuous kilns), and subsequently with the Wellington Machine Company of Wellington, Ohio. As general manager of the latter company, Strickland conceived one of his most successful designs, the Strickland Automatic Brick Machine, in 1920. That machine was one of the earliest of a new generation of high-volume, automatic molded brick machines. He designed and also either built or supervised the construction of a total of ten plants.

The Machinery Makers

Both the Lancaster Iron Works and the Wellington Machine Company designed and built complete brick manufacturing installations, including initial clay processing equipment (granulators, etc.), brick machines, conveyors, and dryer systems. The practice of such design-build programs in the brick machinery business persists to this day. Lancaster Iron Works began as a foundry in the 1850s. The newly immigrated Englishman (that English influence again) Henry Martin, who came from a family of brick machinery makers, joined the firm in the 1860s. Martin eventually made machines in the United States under his own name. His son, James, designed machinery for his father, eventually going back to Lancaster Iron Works as head of the new brick machinery department. Lancaster soon bought up another maker of automatic brick machinery. They also established the Lancaster Brick Company in order to display its equipment in manufacturing brick for a very large new linoleum plant built in Lancaster for the Armstrong Cork Company during the 1920s. Their machines also bore the name 'Posey' due to Posey and Holman's acquisition of Lancaster Iron Works in 1911. Until the mid-1920s, their machines processed the standard nine-brick mold.

The Hutton Company's machine was the latest, the Martin Auto-Brik Model 46 of 1928 that processed a ten-brick mold at the rate of 113,500 bricks per day (the machine was actually rated for 120,000 bricks in an eight-hour day)—which it faithfully did for the next fifty-two years (Figure 58). That such an investment was made in spite of a destabilizing market indicates the commitment of the ownership to the investment program. Even after retirement in 1980, this machine experienced a startling reappearance in the year 2001 because of its high reputation for performance. Typical of the automatic brick machines of the time, this machine automatically sanded the molds, inserted them into the bottom of the press, removed and inverted them for depositing the newly formed bricks onto steel pallets, and returned the molds back to a position from which they could be recirculated for sanding and filling or removed for cleaning. The steel pallets bearing the newly formed bricks were mechanically placed on a cable conveyor that took them to the dryer. The illustrations[10] show a granulator (the metal trench into which the clay cars were dumped, with blades that broke up the clay and moved it toward the conveyor) and the brick machine.

In its essential elements of the vertical mixing box and external press box at the bottom, the new brick machine has striking similarities to Ver Valen's design of 1852 (see Figure 4). At this point, there could be a pictorial summary of the 1920s investment program. A photograph,[11] taken by the maker of the new locomotives, prominently shows the locomotives and dump cars, the roof monitors of the new dryers in the left background, and the new brick machine building on the right. This was the look of the new order—a plant largely rebuilt in seven years—and it was, without exception, a highly successful investment program. The new machines were all electrically driven, and the aggregate power usage required the

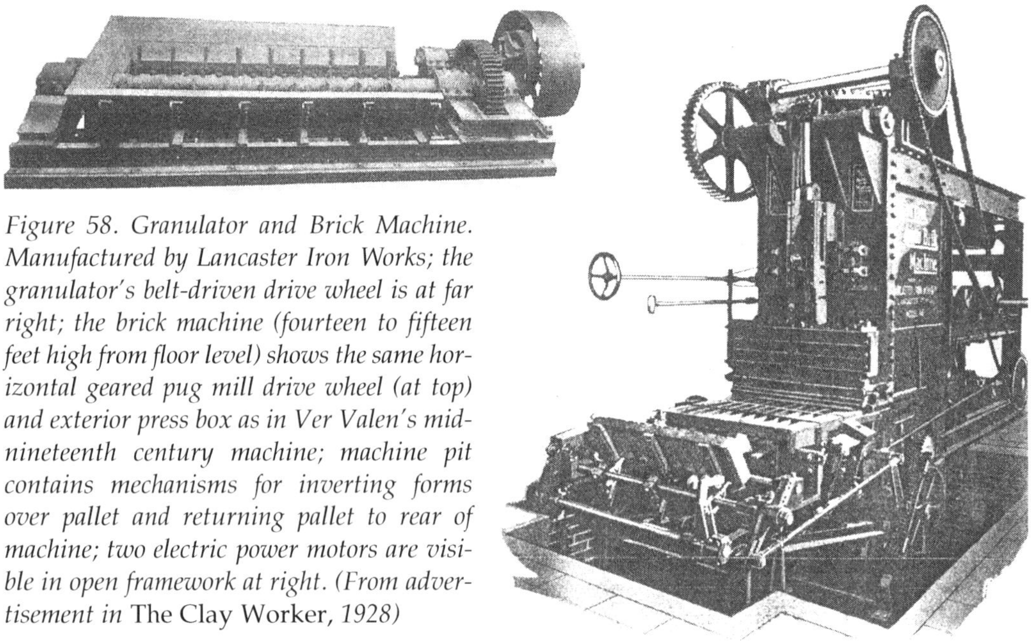

Figure 58. Granulator and Brick Machine. Manufactured by Lancaster Iron Works; the granulator's belt-driven drive wheel is at far right; the brick machine (fourteen to fifteen feet high from floor level) shows the same horizontal geared pug mill drive wheel (at top) and exterior press box as in Ver Valen's mid-nineteenth century machine; machine pit contains mechanisms for inverting forms over pallet and returning pallet to rear of machine; two electric power motors are visible in open framework at right. (From advertisement in The Clay Worker, *1928)*

plant to have its own little substation.

An extraordinary piece of industrial collage struck a whimsical note in the form of a vintage Rolls Royce radiator that was grafted onto the gasoline engine powering the kiln air blowers, proudly bearing a splendid enamel version of the intertwined 'RR' logo on its brass cowl. This same piece of equipment was mounted on a chassis that was supported on wood-spoke automotive 'artillery wheels'—very fashionable in their day—to allow positioning of the blower at the actively firing kiln. Heaven knows if all of this very customized assemblage was the work of some quixotic soul or more probably the result of hard-eyed scavenging. One authoritative source[12] states that this was the work of the inestimable Mont Burhans. The author couldn't take his eyes off of that equipment during his youth, thinking that he had found an exotic hybrid that was unknown to the rest of the world.

Great Efficiencies Suddenly Realized

The ante for staying in the brick business had suddenly gone up, and James Woods' characterization of the brick business from the 1830s ("Brick making is a poor man's game") had become a quaint saying from a distant past. Recall that, before the 1920s—during the times of labor intensive brickmaking with low cost labor—bricks were manhandled individually or in groups of up to six bricks for a total of twelve times. A principal objective in the modernization programs was the extensive elimination of manual brick handling. A limited number of other Hudson River manufacturers aggressively pursued very similar investments in modernization, including Powell and Minnock, Sutton and Suderley, Jova, Dennings Point, Staples, and the Empire Brick Company at Newton Hook (Columbia County). The Jovas, led by the fiery youngest brother, Joe (an RPI engineering graduate), put themselves at the front of this technological revolution, both in terms of time schedule and willingness to try the latest technology—and they were the first to enter into a comprehensive program of modernization, with the possible exception of the more limited modernization of 1920 at the Staples' Malden plant.

The companies making the successful transitions were mostly family-owned firms that had retained the commitment of competent members of the younger generation. However, there were virtually no new entries into the business. In all of this, Van Hutton got a later start and he certainly didn't have the experience of the three Jova brothers with their accumulated years of running the brick business at their plant. GVDH always closely observed and analyzed the Jovas, who were among the first two Hudson River manufacturers to utilize an accessory to the new kiln shed overhead crane capabilities. This was a 'fork' pickup, adopted at the Jova plant in 1924 for the conveyance of 'boat units' (in this case, 1,200 bricks), that allowed crane handling of the units while the bricks were in their three-over-three arrangement required for kiln firing.[13] The bricks were set in the kiln for firing and then carried from the kiln to the barge in that same arrangement.[14] The American Clay Machinery Company first manufactured this equipment, under the inventor's name, as the Penfield Setting System (patented 1908). It had been extensively adopted in the Chicago region. Manual operations for setting the brick in the kiln, and for building the boat units at the kiln after firing, were eliminated—thereby greatly increasing productivity. Another 1920s appearance on the Hudson River of the Penfield Setting System was at the completely modernized Empire plant at Stockport (Columbia County).[15] By the mid-1930s, the Hutton Co. had a related piece of equipment, for loading entire boat units for shipping.[16] Unlike the Jova and Empire systems, which utilized an overhead crane, the Hutton Company system, with its lack of an overhead kiln shed crane (until 1940), required manual building of the boat unit after kiln firing. Before the arrival of the steel sheds, an early type of hydraulic front end lift truck, shown in

Figure 59, facing page, top, left. Overhead Crane with Penfield Setting System Pickup Fork at the Jova Company. (From Hudson River Brick by Brick Manufacturers Association of New York)

Figure 60, facing page, bottom. Overhead Crane on Gantry Loading a Barge at the Jova Company. (From Hudson River Brick by Brick Manufacturers Association of New York)

Figure 61, facing page, top right. Scissor Pickup, 1930s. Suspended from the Lidgerwood crane loading a barge at the Hutton Company; force of gravity acted on scissor leverage to tighten grip on bricks. (Photo by Lipgar Studios)

Figure 62, above. Early Front-End Lift Truck, 1930s. At Hutton Company, taking boat units to crane for loading on barge. (From Hudson River Brick by Brick Manufacturers Association of New York)

Figure 63, below. Bricks Stored Along Dock. Hutton Company; note Lidgerwood crane rail at bottom of brick units. (Photo by Hutton Company)

Figure 62 (in the old Hutton Company kiln shed), took the unit from the kiln shed to the dock crane. Wheeling of the bricks onto the barges with wheelbarrows was thereby eliminated. These kinds of innovations, in a class generically known as 'materials handling,' had great importance in contributing to large productivity increases and continued to be adopted.

The Radically New Tunnel Kiln

The greatest technological plunge had to be the Jovas' installation of the newest development in continuous kilns, two 'tunnel kilns,' the first around 1930 and the second kiln immediately after World War II. Even in the progressive Jova works, the purchase of that first tunnel kiln had to be assiduously promoted by Joe Jova with his older brother, the president, before its purchase. The tunnel kiln made the multichambered Hoffmann type of continuous kiln totally obsolete. The new kilns had a significant introduction in the United States by the early 1920s (one notable eastern U.S. installation for the manufacture of common bricks in Connecticut in 1921), thereby introducing certain manufacturers' names that were prominent into the middle of the century, such as Harrop (Columbus, Ohio) and Dressler, an English firm. Both of those firms ran full-page advertisements on front covers of *Brick And Clay Record* during the 1920s, and both built tunnel kilns at Hudson River plants during the 1950s and 1960s. Ironically, advertisements for both Haigh kilns (Hoffmann type continuous kiln described in Chapter 3) and Dressler tunnel kilns appeared in the same 1918 issue of *Brick And Clay Record*—the old type of continuous kiln competing with its rapidly developing replacement.

Objectively, it must be said that the Hoffmann type of continuous kiln was a hybrid, in that it followed the process of setting and firing bricks in one location. The actual firing moved around the kiln's circuit of adjacent chambers; somewhere in the kiln there was active firing—this being the 'continuous' element. The tunnel kiln's continuous firing stayed in one location, thereby making it more literally 'continuous.' The kiln cars bearing the bricks, move through the firing zone of the kiln. The tunnel kiln appeared in its earliest form in France during the mid-eighteenth century (for small-size porcelain ware), went through development during the nineteenth century, and existed under a number of patents during the early-twentieth century. However, it was used only for light ceramic products like pottery.

Initially, the greatest usage of heavy-duty tunnel kilns was for the manufacture of firebricks (for refractory linings of high-temperature fireboxes, furnaces, etc.). The Jovas' tunnel kilns were typical of the earlier designs. The lack of the post-World War II blower systems caused a troublesome control of temperature. This type of kiln became the absolute standard of the entire industry—and its logic was indisputable. It preserved both the efficiency of operation and the uniformity of firing of the earlier Hoffmann continuous kiln and eventually became far superior in its elimination of the manual handling of bricks. Visually rather dull, it is a long masonry structure through which the bricks pass while stacked on moving kiln cars. Gone is the drama of looking through the scove kiln arches into hell's fires. American plants making extruded brick (with a low water content that allowed stacking before drying), in combination with a tunnel kiln, eventually became marvels of automated continuity, from the brick machine to the completely fired product. The process began at the brick machine itself with the (eventually automated) placement of bricks on the tunnel kiln cars, which continuously carried them through the whole process of drying and firing, thereby eliminating multiple separate handlings—at the dryer and at building into the kiln, as with the scove, Hoffmann, and beehive kilns. The Jovas' tunnel kilns, with their manufacture of soft mud molded brick, still required the manual racking of pallets for steam drying and manual setting of dried bricks on the kiln cars, thereby lacking the full efficiency that was eventually gained with automated handling of extruded stiff mud brick.

In the earlier tunnel kilns, the economy of heat reclamation began at the end of the kiln from which the newly-fired bricks emerge with their heat being used to heat air (for combustion of kiln fuel, similar to Hoffmann-type continuous kilns) that was moving over the newly-fired bricks toward the firing zone of the kiln, with consequent fuel savings. The airflow and the brick cars move in opposite directions. Ducts also move heat from the emerging bricks to the drying and preheat areas, before the bricks enter the firing portion of the kiln.

There was no equivalent to the Hoffmann kilns' extension of heat from actively firing chambers into adjacent unfired chambers. The tunnel kiln is a far simpler continuous kiln system than the multichamber Hoffmann kiln, which required a progression of firing zones around the circuit of chambers, as well as laborious loading and unloading of each chamber. In order to prevent escape of heat into the space below the car decks that supported the bricks (to prevent heat loss and damage to the wheel carriages), the tunnel kiln cars provided a seal at their edges, formed by continuous flanges (integral with the car deck edges) embedded in sand trenches on each side of the cars. This seal was perfected with the use of water in the late-twentieth century. The sand seal feature was invented in 1873. A 1930's tunnel kiln had a capacity of around twenty-four thousand bricks per day[17] (fifty thousand per day according to another reliable source),[18] or about forty-five percent of the net production rate of a scove kiln (one hundred fifteen thousand per day for a one million brick kiln, net, with a liberal ten percent waste deducted). In such a comparison, the soft mud tunnel kiln's greater economy of operation did much to offset its significantly lower firing capacity. The tunnel kiln is continuously fired, twenty-four hours per day, three hundred sixty-five days per year, thereby removing the wear and tear on the kiln structure caused by the Hoffmann kiln's cycles of heating and cooling as the firing passed from chamber to chamber. A recognized problem with tunnel kilns was their vulnerability to damage from prolonged shutdowns and difficulties in restarting after such down periods. Shutdowns were most commonly caused by car derailments within the kiln—a problem that particularly plagued one post-World War II plant.

Van Hutton always looked upon the Jovas' tunnel kiln with studied admiration. He also had respect for the Jovas' willingness to venture into a very new technology—thereby prompting a query as to why an aggressive investor such as Van Hutton did not install such a potentially superior type of kiln. By the end of the 1920s, the investment program had unquestionably used up the available funds for such purposes, and there must also have been a guarded 'wait and see' attitude toward such a new development. Some years later, in 1948, the dean of the New York State College of Ceramics[19] at Alfred University wrote that the use of the tunnel kiln for firing soft mud molded brick, while producing more uniform results than the scove kiln, was not as good as expected. It was not a clear-cut call, considering the expense and the lower production volume from the tunnel kilns existing at that time. Such kilns also underwent enormous improvements during the decades ahead. One highly experienced authority, William Minnock, who had personal familiarity with both tunnel kilns and with scove kilns, states that the earlier tunnel kilns, when firing Hudson River clays, raised testy questions for the risk of capital.

Brick Size Standardization

Note should be made of the 1924 standardization of common brick size by the U.S. Department of Commerce. A large number of brickmakers throughout the country haggled over this issue for many years before the standard was set. The size was 8 inches long, 3-3/4 inches wide and 2-1/4 inches high (thick). Those dimensions were subject to variation due to composition of the clay, differences due to firing, etc. The issue was finally resolved so that everyone using bricks would know that the size was essentially consistent, no matter who made them.

The Largest Modernization of All

The most extreme instance of modernization and renovation upon renovation was that of the Dennings Point Brick Works (operated by the Estate of Homer Ramsdell, its founder), which completely rebuilt its facilities in 1926. David Strickland designed and managed all this until World War II. After two years, there was installed in the new plant a most precocious and radical entry into early automation. This device placed bricks from the dryer on a transfer car in alternate layers of bricks at right angles (bricks in one layer rotated ninety degrees to the layer below—the three-over-three arrangement). The machine 'finger-spaced' the bricks for proper air circulation in the kiln. This new device, under development since 1910, was the Penfield Hacking System (originally designed for extruded bricks). The inventor of the previously mentioned Penfield forklift (Penfield Setting System), R. C. Penfield of New York City, another prolific Hudson River

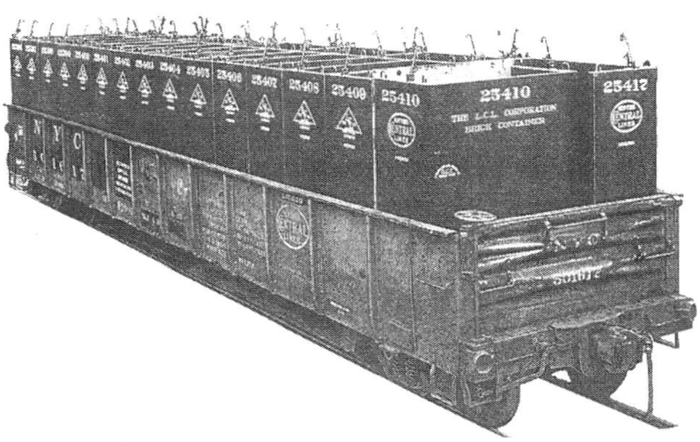

Figures 64 and 65. Rail Shipment Container For Bricks Patented by David Strickland, 1930s. (Photo above by American Car and Foundry Corp; photo right by Pawling and Harnischfeger Corp., both courtesy of Suzanne Strickland)

inventor of clay industry machinery was its designer. Compressed air-operated mechanisms performed lifting functions. Another intriguing new name in the industry had accomplished the actual installation. The New York Brick Handling Corporation of New York City (initially called the Mechanical Brick Handling Corporation of New York) was a specialty firm for the representation, engineering, and installation of brick handling equipment. That firm may very well not only have worked with Penfield in the development and installation of his automated machinery, but, plausibly, Penfield may have been a principal officer in the firm. After installation, the loaded transfer car was then taken to the kiln shed where a Penfield Setting System forklift removed the bricks. The bricks were placed in the kiln, and then placed on the barge after firing by the same means. With this integrated system of materials handling (hacking machine plus crane fork setting), the work of dryer personnel, kiln setters, and boat unit builders after kiln firing was wholly eliminated.

A Sanborn Insurance Map (1927) also shows a feature totally unique to the region's plants—tracks from a rail siding off of the New York Central Railroad's main line that went into the Dennings Point kiln sheds, presumably allowing the shed's overhead cranes to load the railcars.

David Strickland owned patents for containers for brick shipments by rail—which were unquestionably used at this plant. Photographs show a considerable number Strickland's containers in use (see Figures 64 and 65). The new Dennings Point plant had two brick machines and operated as two separate units. One unit operated a standard shift at one hundred thousand brick per day. The second unit ran two shifts per day. Work stopped at 1:00 AM, thereby producing a total of three hundred thousand bricks per day (the recently modernized plants averaged eighty to one hundred fifty thousand per day). Dennings Point's rate of production increased to four hundred thousand per day by double-shifting the other unit. This great production effort was to supply the enormous Parkchester apartment housing project from 1934 to 1938, for which they made the innovative Jumbo brick that David Strickland introduced. Driving the Dennings Point plant at that rate created an enormous depletion of clay resources, located on the east side of the New York Central Railroad's main line tracks. The plant's clay bank railroad went under the main line tracks in a tunnel. Total exhaustion of the clay banks occurred in the late 1930s, causing the plant to be moved to the site of the bankrupt Brockway plant (one and one half miles north of Beacon). Dennings Point purchased it in 1939 and rebuilt with only one brick machine. That plant again underwent extensive change immediately after World War II and in the 1950s.

Another plant to be forced out of business by exhaustion of clay resources before 1940 was the Empire Brick Company of Stockport (Columbia County). That plant had undergone a thorough modernization in 1926, as previously described.

Relentless Change Due to Cement Blocks

By 1927, the number of manufacturers on the river had been reduced by almost fifty percent of the 130 yards of twenty years before. A very good clue to the force behind such a reduction is given in a principal journal of the time, noting the manufacture of 605 million concrete blocks in one year. Insofar as one eight-inch-wide block, used as backup in a wall, displaced twelve bricks, there was a total annual displacement of 7.25 billion bricks, nationwide. This was an ongoing story, with the worst yet to come. In addition, 1928 marked the first time in which concrete blocks and concrete bricks were used throughout a skyscraper, this building in Philadelphia. As a sign of events to come, by 1926 annual American common brick production had dropped twenty-three percent from the 1910 level, and the production of face bricks—a relatively small percentage of total production at the beginning of the century—tripled during the same period, to thirty-three percent of common brick production.

The dominance of face brick in the market was a fact that the Hudson River brick industry eventually had to deal with—after World War II. The year 1926 also saw Hudson River production hit the one billion bricks mark for the last time, and Ulster County was firmly established as the production leader at more than double the production of the next lowest county. Surprisingly, Albany County had the second highest production. Most of this production originated in Coeymans (on the southern edge of the county), with its three plants. Historically, production in the city of Albany had always been consumed in local markets. By 1926, the ranks of the manufacturers increased again, from a low of fifty-two in 1922, to sixty-eight in 1926 (the last big year of the 1920s), only to drop off again to fifty-four by 1929. Those additional sixteen manufacturers, added during the years 1922 to 1926, unquestionably restarted old open yard plants for a short-term crack at the business. The ability of those older yards to lie dormant and re-enter production was a known phenomenon of the business. Soon that phoenix quality came to a sudden end.

Lammies and Clinkers

Worthy of note is the sudden popularity of a type of brick that was considered useless because of damaging deformation due to the proximity to the kilns' firing arches. Bricks in that location would go through an excessively rapid firing that not only caused deformation from rapid moisture removal, but also from insufficient oxygen. This was done to prevent reduction of ferric oxides. The latter were reduced to ferrous oxide, thereby causing bloated, distorted, and fissured shapes with some glassy greenish-black coloring. These were referred to in the trade as 'lammies' (also termed 'clinkers'). Why would anybody want such a brick? The answer must be embedded in the human psyche's restless taste for change. Uniform perfection of shape and color had suddenly become a bore to many people. A fashionable architect's use of such a material could also create immediate interest, these brick having a very picturesque quality. In spite of the difficulties that these brick presented to masons in placing them in the wall, the rustic lammie brick became astonishingly popular during the 1920s. Many Olde English-style houses and apartment houses testify to their popularity to this day. Because of their occurrence in limited quantities from each kiln, they actually fetched a premium price. Clinker bricks also went through a period of great popularity in Germany during the late-1920s, where demand was so extensive that special plants were built for their production.

Cheap Belgian Bricks Invade the New York Market

A mid-1920s influx of bricks from Belgium suddenly knocked down brick prices that had been stable for the first half of the 1920s. Belgium had much lower labor costs and its bricks sold at a price that was fifty percent lower than the domestic product. The American market became the most profitable outlet for Belgian brickmakers. A subsequent reduction in demand of twenty-eight percent for Hudson River brick during the first part of 1927 dropped the price from seventeen dollars to just above twelve dollars per thousand. As a harbinger of decisive events, forty years in the future, the first shipments of brick in rail containers were noted from American plants in climate areas that allowed year-round operations.

Back to Hard Times

With smaller manufacturers only marginally profitable at fourteen dollars per thousand, the Hudson River industry was again facing the same demons of two decades earlier. There was even another effort to consolidate some of the manufacturers, again unsuccessfully. This effort received strongly expressed banking interest, until one of the biggest and best-financed manufacturers persisted in standing aside from any negotiations. The high-cost, unmechanized manufacturers, though operational during the boom years of the 1920s, had a limited lifespan ahead. The modernizing plants were better able to survive the cost-price squeeze, due to an operating cost difference of $2.50 to $3.50 per thousand[20] (eighteen to twenty-five percent of the market price). In the face of deteriorating market conditions, and in recognition of the urgency for the greatest modernization feasible, the Hutton Company proceeded with the purchase of a new central brick machine in 1928, as described above. A manufacturers' meeting in 1927 reportedly discussed whether there was any future in the brick business—by now a familiar frame of mind. Staying in the brick business would take acumen and real tenacity, and only a limited number would succeed. It really was the time of the quick and the dead.

Some New Names

Another name appeared in the 1920s that was noted for effectiveness in promoting the product. Floyd Flint was the secretary of the New York District Common Brick Association. He had subsequently been with the Hudson River Brick Manufacturers Association, and later became general manager and vice president and director of the Hutton Company after World War II.

At the Hutton Company, the mid-1920s saw some passing of the Old Guard, in Pete Lemister's retirement. Sylvester, in his *History of Ulster County,* written in 1880, had mentioned Lemister's name. In 1926, new blood arrived when Thomas Turck joined the company. Turck's Father emigrated from Italy in the first decade of the century. Tom Turck was a very capable man who was responsible for the operation of the brick machine (requiring great care and some of the talents of a magician). He became plant superintendent after William Tierney's retirement in 1953 and later became a director of the company.

Life at the Jova Company

Life around the Jova works was exotic compared to the Anglo-Scotch-Irish ownership at the other plants. Henri Jova's sparkling essay provides memorable glimpses of that world. A hot lunch, prepared by a kindly Hungarian lady was an executive perk. The lunch invariably included canned tomatoes (hot in the winter, cold in the summer), apple pie, and a slab of the esteemed Maillard chocolate. A flow of relatives and visitors from Cuba and Guadeloupe enlivened life at the Danskammer homestead. That house was also the scene of a memorable reception after the wedding of Aunt Tita to Harry Pixley, at which Henri (age approximately ten years) derived a precise ratio for obtaining a full glass of champagne from a given number of almost empty glasses, performing this pioneering computational analysis while hidden under the serving table. A distinguished career as an architect later followed such precocity. This was an extension of the Jova creative genie. Sadly, Danskammer was empty after the death of Juan Jova's wife, in 1928, and that big house fell into disrepair. It had enough antiquity and dark corners that it authenticated cousin Henry Abreu's scary ghost stories, told for the benefit of the grandchildren, about spirit presences from the past (creaking stairs, the guest 'lady in white' when there were no known guests in the house, etc.) within its walls. The great columns from the front are now at the Storm King Art Center. The stones from the walls were salvaged for the construction of a nearby house.

The Worst of Times

Regretfully, the story now turns to the period of the Great Depression, about which almost nothing is said in the Hutton Company directors' minutes beyond noting a drastic 1933 reduction in dividends. The author recollects that GVDH recounted that the plant operated for only two months in 1932—a year that was frightfully harsh indeed. Worse yet, a young worker was killed in the clay banks during the same year—the only known fatality in the history of the plant.

Also occurring in 1932 was an event that could have put the company out of business. The company's excavations had proceeded into the new clay lands that were acquired in 1923, at some distance from the Cordts mansion. However, they had indeed become quite visible from that residence. The residences' owners attempted to obtain a court order enjoining further excavation, first pleading the case in the Ulster County Supreme Court.[21] The plaintiff argued that the excavation was in violation of a 1925 city of Kingston zoning statute. The defendant's attorney argued that the zoning statute was unclear as to limits for the zoned areas. He also argued the merit of the defendant's use of the property having preexisted the city's ordinance. The property had also previously belonged to the Newark Lime and Cement Company, another extractive business. Additionally, it was pointed out that the argued restraint upon the defendant's excavation operations deprived him of beneficial use of his entire manufacturing property, meaning it would shut the plant down. In land use litigation, courts generally avoid the total deprivation of an owner's use of his property. Upon appeal, the arguments by the defendant's attorney, Harry Fleming (his son, Arthur, became secretary of the Department of Health, Education, and Welfare under President Eisenhower), were cited by the two upper courts' referees. The case was decided for the defendant upon appeals by the plaintiff, even though the case was not actually tried before those upper courts. The appeal courts accepted the opinions of the court-appointed referees.[22] The Court of Appeals decision was made in 1934, thereby terminating the case. Even so, GVDH was genuinely fearful for the future of the company, and he somehow always thought that it was a very close call. This case was of at least marginal importance in the development of U.S. zoning law. It had appeared as a footnote in a standard zoning text that the author used in 1954.

Small Pleasures

There was diversion from tough times. The flat land at the top of the bluff to the west of the Hutton Company excavations was turned into a multi-use community activity field, with space for gardens, and a baseball field. It even allowed for two or three families to live there. That field still brings back pleasant memories to men who were boys in the neighborhood at the time.

Moments of Brightness

The construction of the Empire State Building (completed 1931), was one moment of brightness during the depths of the Depression. A number of the river plants made large quantities of bricks, which were delivered to the job at night, as witnessed by the ten-year-old son (William Minnock) of one of the manufacturers. That was one skyscraper job that didn't go to the concrete block producers, and it must have seemed like a miraculous blessing in a time of great need.

More Calamity

In 1933, New York City wholesale brick prices fell to $9.50 per thousand—a drop of thirty-two percent from 1928 and fifty-two percent from the high price of $20.00 per thousand in 1923. This was well below the average break-even price. Needless to say, this was ruinous to the marginal, high-cost manufacturers. It was also a trial for the most competently operated plants, and GVDH's regard for adequate cash reserves and no debt was now the best defense against calamity. Casualties among the brick plants rapidly appeared, and one of the biggest, the Brockway Brick Company (north of Beacon), went under in 1931. Their equipment was bought up by seven of the remaining manufacturers. As previously noted, Dennings Point Brick Works acquired the Brockway property in 1939. Many others also went under, including the big, long-established Excelsior Brick Company of Haverstraw, in 1932. Excelsior's demise, in spite of its new kiln shed with overhead crane brick handling efficiency, and apparently the only fully modernized plant in Haverstraw,[23] must certainly have been attributable to the much higher clay excavation costs (dredging from the river) that typified the later Haverstraw industry. The extent of that modernization, for which they may also have gone into debt, was a curious business decision. It was bloody, and the Haverstraw brick industry virtually ceased to exist after 1936. Older Upriver plants that survived up to World War II included the Shultz and Brigham works, among others.

The End of a Nightmare

The year 1936 brought the beginning of the end of the nightmare. Production thereafter gradually resumed, but not rapidly enough for real optimism. GVDH gave a pessimistic report in 1938, and a subsequent directors' minutes note from 1939 stated "considerable promise, large volume. . .modest profit."

Complete Unionization

In 1938, the Brickhandlers Union of the International Longshoremen's Association (Local No. 1467) unionized the Hudson River brick industry. This association—often troubled—lasted until 1966. This was done on the crest of the organizational militancy of the American labor movement of the 1930s, with assistance from the collective bargaining provision of the Wagner Act of 1935—and at a time when the Hudson River brick industry was just barely viable again. The association of a stevedore's union with brick handlers may well have originated with organized stevedores who forcibly curtailed the unloading of barges at city berths by live-aboard brick company crews in 1910 (see Chapter 2). The International Union of Operating Engineers, Local No. 825, represented equipment operators and maintenance men. The barge captains had a separate union until they eventually merged with the Teamsters Union. Although relations with the unions could be troublesome and a cause of increased operating expenses, Van Hutton gave unqualified credit to the trade union movement, nationwide, for significantly raising the standard of living of American workers. That was justification enough. The combination of increased productivity from investments in efficient production facilities, and the ability of the manufacturers to raise prices for a number of years after World War II allowed profitable operations, thereby offsetting increased labor costs.

The Parkchester Project

The year 1938 saw the completion of construction of an enormous apartment project, commenced in 1934, that the Metropolitan Life Insurance Company financed, designed, and owned (Figure 66). This was the very notable Parkchester complex in the East Bronx that was without precedent in the sheer volume of bricks for a single project. The Parkchester project, at over twelve thousand apartments, was an extraordinary accomplishment that concealed its great size by artful organization and composition, including provision of stores, offices, a sizeable theater, and a convenient subway stop (elevated service at this point) for access to Manhattan and to the nearby Pelham Bay parks. Literally 'artful' was the inclusion of ceramic sculpture figures on building corners (at various heights) and elegant ceramic panels over entries depicting various animals.[24]

figure 66. The Parkchester Apartments. A small portion of that large complex. (From "Hudson River Brick by Brick Manufacturers Association of New York")

Besides being built with Hudson River brick, the appearance of a radically new type of brick—the

Jumbo—marked the project. The perpetually inventive David Strickland conceived this idea, as previously stated. The innovation was a simple increase of the brick height from 2-1/4 to 2-5/8 inches, thus increasing the wall area occupied by the single brick and decreasing the time required for laying a given area of wall surface. Eventually, the masons' unions set a limit on the number of Jumbo bricks to be laid in a day. However, the Jumbo brick dominated the industry after World War II. With the deep setback of the window frames from the outer wall surface, the Parkchester wall construction appears to be of the massive masonry type, which means enormous quantities of bricks were used.

Other Projects of the 1930s

Notable projects containing Hutton Company bricks during this period included Meadowlawn, a handsome residence on a sizeable Riverdale estate, now owned by a private public service organization. Hutton Company bricks also were used in the construction of the Cloisters Museum and what must have been a good number of Park Avenue apartment houses. A recent occupant observed HUTTON brand bricks being removed from an interior wall during alteration work. Other notable Hudson River brick projects from the 1930s included a low-rise apartment project that the preeminent planner Clarence Stein designed, as well as several handsome swimming pool complexes for the New York City Department of Parks under Robert Moses. Completed projects also included the Hayden Planetarium and a number of other publicly and privately financed apartment complexes.

A 'Pre-Owned' Steel Kiln Shed

Final elements of the Hutton Company plant modernization program were completed in 1940, the most principal of which was the direct result of increased labor costs incurred with the unionization of brickhandlers. This was the installation of a 'pre-owned' steel kiln shed, with a new overhead crane[25] and a new gantry that cantilevered out over the river for barge loading.[26] The aerial view in Figure 67, though taken in 1948, shows the plant modernization, which was completed by World War II. The ends of the kiln shed's overhead crane were supported on wheeled trucks that rode on rails supported on columns on each side of the three main kiln spaces (bays). The rails extended onto the cantilevered gantry to enable the crane to position itself over a barge. This allowed unlimited access to the kiln areas as well as the barge decks. Narrower spaces between the large kiln bays were for access, fuel lines, and blower ducts. The appearance of this essential facility at the Hutton Company was the result of another plant's misfortune—the 1932 bankruptcy of the Excelsior Brick Company of Haverstraw, which had erected the shed, brand new, in 1928. The Bergen Point Iron Works (Bayonne, New Jersey) erected that structure to the design and specifications of the New York Brick Handling Corporation—previously mentioned as the installer of a radical automated hacking machine at Dennings Point in 1927. The New York Brick Handling

Figure 67. The Hutton Company, 1948. The completed modernization program, including the steel kiln sheds erected in1940 (lower right), with gantry at right-hand kiln bay. (Photo by Air Photos Associates Inc.)

Corporation unquestionably engineered and specified the size of the shed's kiln bays, the intervening service aisles, arrangement of crane rails, roof vents, structure, etc. A number of such big sheds were built for the modernizing brickmakers—and it can be reasonably conjectured that they had the same origins (the New York Brick Handling Corporation and Bergen Point Iron Works).

The Excelsior sheds were dismantled, loaded on a barge, and moved Upriver—a bargain at fourteen thousand dollars. Recall a 1926 authorization by the Hutton Company board of directors to purchase the same facilities for a hundred thousand dollars. This amount was undoubtedly based on an accurate estimate—although this estimate must have included a gantry and an overhead crane. The latter items may have been separately sold after Excelsior's bankruptcy. Van Hutton loved a real bargain, and the Excelsior pre-owned shed—a dozen years old—was a delectable bargain indeed. New accessories for the sheds included a gantry (also by Bergen Point Iron Works) and a new P&H overhead crane (Pawling and Harnischfeger Corporation, Milwaukee, Wisconsin). The gantry was rail-mounted in order to allow positioning at the kiln shed bay, which contained fired bricks ready for shipment. The immediate result of the kiln-shed installation was to reduce the workforce by approximately fifty workers. This kiln shed represented some kind of commitment to the future use of scove kilns, even with Van Hutton's knowledge of the Jovas' tunnel kiln. In spite of the incremental nature of the modernization over a period of seventeen years, one critical peer judged the final result to be a very compact and efficient assembly of parts (Figure 67). After the adoption of the tunnel kilns, which were housed in separate structures, the Jovas continued to use their big scove kiln sheds for storage of bricks that were made late in the season.

Company Barges

In addition to the shed, three new barges were built for the Hutton Company at

Figure 68. Hutton Company Barge. Being towed through company's channel. (Photo by Hutton Company)

the nearby Hildebrandt yard. Barges were named after men of distinguished service to the company. The *Englebrecht* was named after one of the esteemed barge captains, and the *Tierney* named after the superintendent who had been at the plant since 1906. The third new barge was named after the company's cofounder (or very possibly after the author's older brother), the *William Hutton*.

The plant's forces did a great deal of the maintenance work on the barges, caulking deck seams with oakum (a coarse hemp or jute fiber loosely twisted into ropes). These were hammered in place with specially designed caulking hammers (they looked somewhat like a croquet mallet with a very elongated head with wide metal bands around it) and caulking irons (blunt, chisel-shaped tools) for compacting the oakum into the joint. All of this was sealed into the joint with hot pitch. The plant's personnel worked from a float on vertical hull surfaces above the (unloaded) water line. Repairs below the water line had to be done at the Hildebrandt yard. That yard had a floating dry-dock, a broad-deck vessel that was sunk in order to allow the repair boat to ride over it. The water was then pumped out of the floating dry dock, thereby lifting the repair boat out of the water. The staccato "tock! tock! tock!" of the Hildebrandt shipyard's caulking hammers was the most memorably unique audio-trademark of that part of the Rondout Creek—very distinctly heard from the (former Route 9W) suspension bridge, and from a large part of Rondout. Hildebrandt continued to build big, heavily framed vessels of wood right through the Korean War, making 140-foot nonmagnetic minesweepers (using lots of non-ferrous fittings and fastenings) at that time, as they did during World War II. One of those ships, a survivor of the disastrous typhoon at the end of World War II, which caused more damage to the U.S. fleet than any combat action of the war, was seen in the petroleum exploration service in the Gulf of Mexico during the late 1940s. It had preserved a steering eccentricity acquired from its experience in that typhoon.

The author witnessed a couple of Hildebrandt's skilled adze men in the act of perfectly shaping a massive keel section with a varying taper along its length. That piece hardly needed additional finishing on its surfaces. Barges could be severely damaged at their berths in Manhattan, as happened in the Harlem River when a passing vessel sideswiped a Hutton Company barge, causing a good portion of its load to be violently dumped into the water and badly holing the barge.

A Memorable Voyage

Even though the following events happened right after the end of World War II, the author will use this discussion about barges to recount an absolutely unforgettable three-day voyage to Manhattan on a Hutton Company barge. The barges had a single cabin at one end for the captain, who, in this case, was the highly respected Captain Morgan. A small Cornell service tug (the *Rob*) took the barge out to 'stake piles,' to the south of the mouth of the Rondout Creek. It was tied up there to await the big 'tow' as it came down the river. Since the late-nineteenth century, the Cornell Steamboat Company had done all of Hutton Company's towing. Actually, there was no choice in the matter of towing services, since Cornell had a monopoly. Upon arrival of the southbound tow, its large tug cast off in the river channel and proceeded over to the waiting Hutton Company barge, taking it back to the tow. There were probably fifteen to eighteen barges in the complete tow, carrying bricks and trap rock. Some barges had live-aboard wives, geranium boxes, pets, etc.—and all very sociable.

Captain Morgan immediately made it very plain that he hated to cook, gladly eating whatever the author prepared, no matter how superficially unsavory (probably comprised of Spam, eggs, spaghetti, canned vegetables, and cabbage for cole slaw). The cuisine was prepared on a kerosene stove, with a limited amount of very rustic, self-taught campfire food preparation as culinary apprenticeship for the *chef du maison*. Captain Morgan himself pronounced the fare quite tasty and was particularly grateful for not having to cook. In order to make that point clear, Captain Morgan insistently remarked on the author's enthusiasm for cooking ("You like to cook, don't you, George?").

Opposite the Palisades, a second big Cornell tug assisted the tow. The tugs were abreast of each other (their names *Lion* and *Jumbo* in big white letters on their low, black sterns), and those two churning brutes fought a losing battle against the incoming tide—we were going northward for a while. After admiring the city's lights after nightfall, the author hit the hay. Around 3:00 A.M., the author awoke, hearing a lot of crashing around on deck and the harsh metallic rattle of the barge's capstan (a vertical deck winch for securing tow and dock

figure 69. Brick Crusher at Hutton Company. This machine produced pulverized material for surfacing the base lines at Yankee Stadium. (Photo by Hutton Company)

lines) directly on the other side of the cabin wall by the bunk. Hurriedly turning out, the author found the barge tied against the side of a small harbor tug, proceeding around the Battery toward the East River. Jumping aboard the tug, an open door revealed a fireman (quite affable) who was hand-stoking the tug's boiler, methodically opening the two fire box doors with the edge of his shovel and flinging in big scoop-shovel loads of coal into the glowing fireboxes. The barge was finally berthed in the East River at lower Manhattan, thereby concluding a splendid little adventure.

Family Matters

In the period around 1940, there were some other family business anecdotes worth recounting, such as the stockholder who sought to elicit support for raiding the healthy contingency reserve that GVDH steadfastly maintained, for the purpose of substantially increasing dividends. There was no support in this matter from any of the other stockholders who respected the president's high regard for the fiscal prudence that had just pulled the plant through the Depression, transformed the plant into a modern manufacturer, and made it one of the leaders in the industry.

A New Rationality

Van Hutton, along with such contemporaries as Joe Jova, was a strong believer in the hardheaded analysis of methods to improve the product. To this end, there was established, in the late-1930s, an initial relationship between the Hudson River brick industry and the New York State College of Ceramics, at Alfred University, which was the center for the Ceramic Association of New York. Its Ceramics Experiment Station (a laboratory) commenced operations in 1936. The State of New York subsidized this research work, which was significant for its extensive involvement in technical matters affecting the brick industry, as well as other types of ceramics manufacturing.

It was now possible for the manufacturers to have highly qualified technical assistance in solving problems that were specific to Hudson River brickmaking. Subjects addressed by the laboratory included new national standards for water absorption that would reappear as troublesome topics after World War II. The acuity of the examination by the laboratory of the applicability—or inapplicability—of those standards is impressive, and they foreshadow direful events to come. There was other evidence of cooperation between the Hutton Company

Figure 70. Group Portrait at the Hutton Company,1940. Back row (left to right): Frank Perry, Joseph ('Pepe') Naccarato, Pete Lemister, Sam Frangello, Gene Frangello. Front row: Salvatore (Tutti) Rienzo, Clayton Stalter, Van Hutton, Mike Naccarato. ("Oh, why did the boss want to take this picture in the dead of the winter? Why not in the summer, when we're thinking about baseball and our gardens?")

and Alfred University, as indicated by a ceramic engineering thesis student (in 1938) who acknowledged the assistance of William Tierney (the plant superintendent) and GVDH. A much closer relationship between the manufacturers and the College of Ceramics at Alfred University commenced a decade later.

The Christmas Club

A savings plan for Hutton Company employees, the Christmas Club, was established in 1940 in cooperation with a local savings bank. The company paid ten percent on accounts in the savings plan, and all money in the accounts was delivered to employee-depositors before the holidays, for Christmas or any other needed expenses.

Termination of Production by World War II

Insofar as brickmaking was deemed nonessential to the war effort, the plant was shut down during World War II, retaining a maintenance crew and doing improvements toward the re-opening. The rest of the men were in the armed services or working on the construction of Stewart Field, at Newburgh. GVDH took a position with the War Department in New York City, renegotiating military contracts with contractors who seemed to be making too much money off the war. He took weekend trips to the brick plant in order to keep an eye on things and to develop strategies for the postwar startup. It was also during this period of time that the first non-family members were appointed to the board of directors—William Tierney (plant superintendent) in 1943 and Montgomery Burhans in 1945.

After World War II, Hudson River brickmaking came back to life, with the survival of only twelve plants.

Notes

1 Williams College graduation photograph–Figure 51.

2 Adjusted for seventy percent inflation 1915-1923 (per U.S. Dept. of Labor CPI), this price is around $11.80 in 1915 dollars.

3 Holland House—photograph late-1920s—Figure 52.

4 Figure 53—Steam dryer showing cable conveyor in middle and pipe racks on each side.

5 Lidgerwood crane—Figure 57.

6 McCullough. *The Path Between the Seas.*

7 Chernow. *The House of Morgan.*

8 Francis Duffy (see Bibliography) verifies the existence of the dual granulators at the Hutton Co. when he joined in 1933.

9 See APPENDIX—Mixture Control and Additives.

10 Figure 58. Showing granulator and brick machine.

11 Figure 54. Photo by Vulcan Foundry.

12 John Polacco (see Bibliography).
13 Figure 59. Overhead crane with brick unit hoist.
14 Figure 60. Gantry, with crane loading brick unit on barge.
15 *Clay-Worker*. Dec. 1928.
16 Figure 61. Boat unit scissor-compression type of pickup.
17 Dobson. *A Rudimentary Treatise on the Manufacture of Bricks and Tiles.*
18 *Brick and Clay Record*. Oct. 1927.
19 J. F. McMahon, author of *A Critical Study of the Manufacture of Soft Mud Brick.*
20 *Brick and Clay Record*. Nov. 1927.
21 Cordts vs. Hutton et al.—146 Misc. 10, Ulster County Supreme Court 1932.
22 _____. 241 AD 648 Appelate Division 1934 and 266 NY 399 Court of Appeals and 241 NY 400 Court of Appeals 1934.
23 Sanborn Insurance Map of Haverstraw 1931.
24 Jacoby. *Architectural Sculpture in New York City*. Figures 60 and 62, the latter is a corner figure of a woman with a fully deployed umbrella and an enigmatic pursed-lip expression (of surprise or possibly reproach—or other choices left to the viewer's puzzlement).
25 Similar to Figure 59.
26 Similar to Figure 60.

Principal Sources for Chapter 4

Brick and Clay Record. Feb., July, Aug. 1927; Jan. 1928.
Brick Manufacturers Association of New York. *Hudson River Common Brick* (two volumes).
Ceramics Experiment Station. (Alfred University) *Monthly Reports.*
Clay-Worker. June 1928
Duffy, Francis. (See Bibliography.)
Engineering News Record. *Construction Costs.*
(Engineering documents for Hutton Company equipment—in collection of Hudson River Maritime Museum).
Hutton Company. *Minutes of the Stockholders and Board of Directors.*
Loose, J. *Notes on Lancaster Iron Works.*
Museum Bulletins of New York State. *Mining and Quarry* series. #273, #277, #295, #306, #319.

CHAPTER 5
A Last Round Good Times—1945-1956

THE BEGINNING OF THE DECADE following World War II was a time of very optimistic expectations for the industry, even though there were only a dozen plants in operation at that moment. Approximately half of those were genuinely well equipped and competently led. Those ten years ranked among the best in the entire history of the industry, in terms of brick quality, productivity, general profitability, and enlightenment among the manufacturers. The rapid decline of the remains of the Hudson River brick industry that had endured past World War II followed. Those years were the last hurrah.

One Plant's Experience

Van Hutton's 1945 assessment of the volume of business in the postwar period was sufficient enough to warrant the establishment of a new corporate officer at the plant—general manager—with Floyd Flint (who represented the Hudson River plants in the city during the 1920s and 1930s) admirably filling that position until the sale of the plant in 1965, at which time he was vice president. Highly personable, Floyd Flint was an asset—as well as a lover of good cigars, an excellent golfer, and a memorable man. Established lore at the plant had it that he had once been on the short list of tryouts for third baseman with the Cleveland Indians. On Flint's arrival in 1946, GVDH said, "Floyd, we're going to beat the devil himself. . . ."—and they did. The blood was up, it was a splendid time to be in the brick business, and they were still young enough to hit the line very hard and enjoy it. There was a pent-up demand for housing construction, fed by over fifteen years of neglect due to the Depression and World War II—and this was clearly understood.

Fine Tuning the Scove Kiln

The Hutton plant began postwar production with upgrades to the scove kiln installation that would maximize the productivity of that type of kiln. They added permanent brick floors (traditionally, kilns were built on dirt floors that had to be accurately graded) and permanent kiln walls (on three sides, thereby eliminating the mud daubing that was needed for air tightness and heat retention, with thirty-one arch openings on each side). A lot of that work had been done with the maintenance workforce during the war years, and the author and his brother helped to lay down an awful lot of brick paving. The scove kiln was to be pushed to the limit of its productive capabilities, and the plant's kiln personnel had the ability to do just that. Scove kiln waste was down to six to seven percent, half of which was the result of the 'double coal' bricks—an irreducible part of scove kiln operation. Actual waste from kiln burning—unusable bricks due to insufficient or excessive burning—was therefore almost in the same range as the waste from a tunnel kiln (two and a half to three percent).

The introduction of the 'Priester fork' (Bradney Machine Company, Middletown, New York) greatly improved kiln shed efficiency. It picked up units of dried brick that had been manually stacked into three-sided transfer cars at the driers.[1] This device certainly had considerable similarity to the Penfield Setting System fork that had been developed earlier in the century. Manual handling of bricks was thereby reduced to that one operation at the dryer. The Priester fork was then used to put the units into the kilns for firing and subsequently to load the fired units onto barges.

Investment as Continuing Focus

The possibilities for significant investment received intensive study over the coming decade—and the will to invest was very strong indeed. Lest there be any question of the use of significant retained funds for investment purposes (recalling a 1940 stockholder's challenge to such retention, in recommending surplus

Figure 71, facing page. Firing a Kiln at Former Hutton Company, Late 1970s. The oil atomizer nozzle plainly visible at right arch, control handles for oil supply, air duct (on floor), and permanent kiln wall; heavy canvas curtain at right is to prevent excessive air movement toward kiln arches. (Photo by Tripoli)

Figure 72, above. A Crane Operator's View. Priester fork at Hutton Company taking 'unit' from three-sided car that has been brought to kiln shed (by a 'transfer railway') from a dryer, where the dried bricks had been assembled into a 'unit' in the car. (Photo by Hutton Company)

distribution to the stockholders), a 1946 supplement to the directors' minutes was entitled "Why We Deem It Advisable to Retain a Large Part of Net Profits Rather Than Distribute Them"—an analysis of targets for capital funds. There is no mention of a tunnel kiln, that decision was made self-evident before 1946 in all of the scove kiln upgrade work on the permanent kiln walls—and this very critical and intriguing decision will receive further consideration herein.

Initially, the dryers became objects of scrutiny because of a sudden increase in the labor costs of that part of the manufacturing process. Manual hoisting of the brick pallets from the conveyor to the pipe racks was hard work–and rackers' wages had become among the highest in the plant. In addition, the Brickhandlers Union had negotiated for an additional racker to be added to each shift. One well-informed account recently indicated that this arrangement might have been unique to the Hutton Company. A somewhat troublesome defect in the dryers' operation, due to the incremental modernization of the 1920s, also drew attention. That defect was in the dryer pipes, which were arranged for a nine-brick mold (the standard mold at the time of the dryer construction from 1922 to 1925). The 1928 LIW Martin brick machine (as purchased by the Hutton Company) processed a ten-brick mold—causing uneven drying of one brick on each pallet. The purchase of the brick machine had to be a calculated risk for obtaining greater productivity—and the misfit of the ten-brick molds with the dryer racks was a bearable nuisance. Therefore, a survey of alternative drying methods—focusing on 'car dryers'—became a priority, in addition to the purchase of various equipment items (including two diesel power shovels, an additional kiln shed crane, and the possible addition of a fourth bay to the kiln shed).

Two Hudson River plants had new dryer technology installations (including one example by the still very inventive David Strickland), and GVDH made very plain his critical analysis of the operation of those installations. One advanced dryer system under consideration used jets of hot air for drying. Other companies' experience with new methods and equipment always received his very close scrutiny. Included in the study was a summary of the condition of critical pieces of equipment in the plant, with comment on their age. In all, this was a careful analysis of the plant's condition—the kind of intellectual exercise that was greatly appealing to GVDH—and the total cost of this program was substantial. The study of 1946 was only an introduction to many updates on this subject, to be pursued over the next several years. Immediate purchases in 1946 included new railroad equipment (Plymouth diesel locomotive) and a Caterpillar bulldozer. The resolution, in 1948, of a fully-funded "Long-Term Capital Asset Replacement Program of Five to Ten Years Duration" firmly established the intention of this investment program. It was to have annual increments made to the initial amount, with its own entry on the account ledgers. Significant investment was to take place. Having known the dark side of the brick business and the Great Depression, investment was again to be accomplished with retained earnings—cash generated by a very active market.

Figure 73. Essential Men at Hutton Company. Montgomery ('Mont') Burhans and John Sepesy. (Photo by Hutton Company—also in dead of winter)

Talented People

At that moment, the great talents of Mont Burhans and the plant's own forces were powerfully displayed in the erection of an additional kiln shed crane (manufactured by Whiting of Harvey, Illinois). Two cranes were required, with one crane to build up a kiln for firing and the other to take down a fired kiln for transfer to a barge for shipment. The older, slower P&H crane was used for building the kilns, insofar as it traveled a shorter distance than the faster Whiting crane. Its additional travel distance entailed going out onto the gantry for barge loading. Rather than hiring a rigging firm, plant personnel used a 'gin pole' derrick to swing the machine beam into place. It spanned the forty-five-foot-wide kiln bay—riding on rails almost thirty feet above the kiln floor. Erected by riggers, the P&H crane installation of 1940 set a precedent. The control cab-hoist unit, riding on the machine beam,[2] was then hoisted into position on the beam. Mont Burhans directed the whole procedure, which the plant's personnel admirably executed. These were ungainly, very heavy assemblies of metal requiring careful handling and precise positioning on the rails—a real feat that is usually performed by pricey professional riggers.

A similar instance, shortly thereafter, was the installation of an extension to the south granulator trench and its machinery. An engineering firm put a high price on this item, whereupon Mont Burhans took over the job with complete confidence, directing its very successful fabrication in the plant's machine shop. Chambers Brothers of Philadelphia furnished the machinery parts for this granulator. This company was one and the same as founded by Cyrus Chambers, who pioneered the extruded brick machine of the 1850s (becoming Chambers, Brother and Company at that time). They were still very much in business in the mid-twentieth century. The plant developed superb maintenance personnel, as will be seen, and they had many opportunities to apply their talents, which were honed on a regular basis during the annual winter shutdown.

January and February were devoted to the dismantling of the brick machine for its cleaning and replacement, or in-house rehabilitation, of parts. The men became experts at building up the metal surfaces of worn parts and machining them to fit into the brick machine. These same skills were also applied to other equipment. Rigorous maintenance—highly valued by GVDH—was developed

Figure 74. The Powerful Graphics of the East End of the Kiln Shed, Late 1970s. The crane control cab; gantry framing is visible at lower left of opening. (Photo at former Hutton Company by Tripoli)

into a fine art, eventually prolonging the life of equipment well beyond normal life spans. One of the men of the old regime, who could see the price paid for the neglect of stringent care, severely criticized a post-1965 successor owner who did not follow the same discipline.

Stuytown and Peter Cooper Village

The year 1947 saw the beginning of several of the most eventful years in this plant's long history. A steep but brief 1945 national recession following World War II introduced a period of prodigious building activity in the New York market, including the beginning of the construction of Stuyvesant Town (affectionately known over the years as Stuytown), a mammoth housing complex that would have a fascinating history in the years to come and in its projection into the third millennium (Figure 75). Metropolitan Life Insurance Company financed and owned this project. They were the same builder-developer that did the Parkchester complex of the 1930s, and both projects were built by the same contractor (Starret Brothers). Stuytown extends from Fourteenth Street to Twentieth Street and between First Avenue and FDR Drive at the East River. Upon completion of this project, a smaller (by approximately fifty percent) increment immediately followed—Peter Cooper Village—extending the complex to Twenty-third Street and providing apartments, in both complexes, for a total of twenty-five to thirty thousand people (in 11,250 apartments) on eighty acres of land. Those two projects were phenomenally successful demonstrations of middle class urban living, and they will get further attention toward the end of this writing. Their construction used vast quantities of Hudson River brick—and the industry, including every employee to a man—could take great pride in having furnished the most essential kind of building material to such a splendid undertaking.

Even though these two projects—and Parkchester as well—were built after "The Golden Age" of the industry at the beginning of the century, they had to be unprecedented at any time in the history of the industry for sheer size and consumption of bricks on single projects. All of those brick were of the Jumbo size that had been used at Metropolitan's Parkchester complex before World War II, and that type of brick dominated the market for the next two decades.

Importantly, Stuyvesant Town characterized the new reality of the brick business—all bricks would henceforth have to be 'face quality' brick with minimal imperfections. That project used brick for exterior finish (brick veneer) purposes only. The full implications of this new reality became apparent within a few years. Gone forever were the days of massive masonry bearing walls requiring the huge quantities of bricks of various levels of quality that had been the life blood of the Hudson River brick industry. Nevertheless, it was still a very big market. A ceramic engineering authority (Wayne Brownell), in a 1950 publication by the New York State Department of Commerce,[3] gives a notion of the sheer quantity of bricks manufactured at this time, as well as the productivity of the mid-twentieth century Hudson River plants. Brownell's writing states that the

Figure 75. Stuyvesant Town Apartments. A small portion of that project; note complex plan shape of buildings, with many outside corners for obtaining very desirable cross-ventilation, in pre- air conditioning era. (From "Hudson River Brick by Brick Manufacturers Association of New York")

Hudson River plants made five hundred million bricks per year—almost forty percent of the greatest production in the history of the industry (in 1906). Furthermore, ten percent of the number of plants existing at that earlier time accomplished this. This is one indication of the productivity increase accomplished by the earlier modernization programs.

New Accounts in a Roaring Market

A pell-mell brick market characterized those post-World War II years, with buyers literally pounding on the plant's door, so great was the need. Big New York accounts, mostly very new, soaked up great quantities of bricks, including Brooklyn Builders Supply (Bud Buschmann), Arrow Builders Supply (Steve LaSalla), and General Builders (Jim Farley and Samuel Schiff). Buyers' demand for direct purchase from the manufacturers jeopardized the services of the traditional commission men brokers, such as Barnes, McNamara, and Morrissey. However, that firm—which had certainly recognized the Van Hutton of the early 1920s as newcomer talent that should be well regarded and assisted in prospering in the business—commanded durable loyalty from the Hutton Company president, which persisted in the face of the roaring new market. GVDH attested to this in a remark made to the author's brother: " Everyone gets a chance to walk over someone in their lives—and we could have done so in this instance—but we didn't." Old loyalties held fast and the relationship continued.

A visit to the office of Barnes, McNamara, and Morrisey (in the Grand Central Office Building), during the late 1940s, revealed a splendid array of varnished wood rolltop desks and a fine, mellow scent of long-time cigar smoking (absent spittoons)—a scene with an imperiled existence. They could still be of real assistance in locating barge berths and in providing in-town servicing of accounts. That did not relieve GVDH of his weekly visits to the city that might include personal inspections of moored barges in places that could be a bit unwholesome, like Brooklyn's Gowanus Canal (a fall into that poisonous water could kill you). Barge captains were also required to send daily postcards back to the plant, stating the number of brick remaining on the barge, etc., for use in disputing 'demurrage' charges for occupying a berth beyond a contracted period of time. An enormous accumulation of those cards had piled up in a battery of cubbyholes on one wall in the office—quite unexplainable in such a tightly run operation (they may have been considered decorative, in a business way). Other very big apartment projects followed, such as the Dyckman Houses. The Hudson River brick industry was in heaven, for a change, after having been through hell for so long.

Do-It-Yourself Dredging

The shipping operations of the business also included maintenance of a lengthy company channel from the main river channel through the shallows (four feet deep at low tide) to the plant's dock. This became problematical when the Cornell Steamboat Company went out of business. Its small local tug had towed the loaded barges out to the river. Callanan of Albany took over towing. Callanan's big tow tugs cast off their tow in mid-river to come in after the barges through a channel that was marginally large enough, and which their passage disrupted (collapsing the sides of the channel). Small dredging firms had ceased to exist on the river, with only very large harbor dredges available. An improvised solution was devised for this situation also. The company's dragline crane was placed on one of its own barges. The Hutton Company was in the do-it-yourself dredging business.

A Narrowing Market

Observations at the end of 1948 broadly hinted at the future of the brick market—a drastic narrowing of the customer base. Single family home construction that previously used appreciable amounts of brick suddenly became priced out of the individual house market by high labor costs for masons—leaving the brick market to be dominated by big apartment projects, largely government funded. Such a narrowing of the market was very formally noted in the directors' minutes of 1948, with its long-term implications plainly understood. Additionally, an expanding network of subways, commuter rail service and, most importantly, new systems of express road construction that enabled more remote commuter living (Nassau County population doubled in the 1950s to more than one million, and Suffolk County followed suit in the 1960s) rapidly dispersed the formerly compact metropolitan New York market.

Life at the Plant

It was at this time in the later 1940s that the author spent his last summer (1948) working at the plant—and forming some indelible memories. Principal among these was the observation of GVDH's management style. This can be depicted in a situation where he is standing with two or three men, in animated discussion, vigorously eliciting their opinions and doing much listening. He loved both using his own mind and engaging with others who could do their own thinking—and those men's valued contributions were highly beneficial to plant operations. The author saw many of those same vignettes during that summer—and they were evidence of Van Hutton's admiration and respect for so many of the men who did their work so well and with such good spirit. These men were also fine family men—an attribute that GVDH held in high regard. Great value was placed on loyalty, with full knowledge that this quality was very much of a two-way street.

The senatorial and very competent Charles Roth (at the plant since 1923) controlled the demeanor of the plant's office—no all-week TGIF in there. That office was hard-edge and austere, as exampled by the author's suggestion to GVDH that a couple of window air conditioning units would allow the office workers to avoid working in their undershirts, with a big fan blowing hot summer air around the room. That suggestion was scornfully received, with the dismissive statement that, "If any man from the machine room or kiln shed came into a nice, cool office, we'd have instant labor problems." The author's work kept him in the clay banks during the mornings, where he had eye-rubbing daily sightings of the southbound dayboat gliding in for an 11:45 AM landing at Kingston Point while emitting a beautifully wailing (if slightly hoarse) whistle. Most spectacular of all dayboats was the *Robert Fulton* (decommissioned in 1954), a living dinosaur, with its archaic (and highly visible) walking beam and churning sidewheel paddles. Ironically, that was the last season for such visions. The New York-to-Albany service ceased at the end of that same summer. The afternoon was spent in the machine shop, which often included assisting in tasks like fastening the protective iron wear plates to the brick molds. On one such occasion, I was adjacent to John Heppner—who could only be described as a splendid fellow and a manic worker. John was doing the same job and the author thought that his furious pace, with a large, heavy-duty electric screwdriver, was the exemplar to be followed. In a while, there was a hand on the shoulder and a low voice saying, "You don't have to try to keep up with him—none of us can." As I remember, that voice belonged to the avuncular, rock-solid John Sepesy.

Lunchtime in the shop, attended by most of the operating engineers, had a remarkable formality. Mont Burhans kept a folding wood chair against the shop wall that he deployed in the middle of the shop floor every day at noon, there to *preside* (italics for emphasis) over lunch, with the men sitting on the workbenches around the walls (a highly socialized gathering of good men). Mont had a hierarchical view of his world—and in that world he occupied the apex of the pyramid. Baseball was always a safe but vigorous lunchtime discussion topic, and certain workers had well-known loyalties. On one occasion, in which a big fan of the newly-minted superstar Jackie Robinson appeared in the shop door simultaneously with a distant clap of thunder, one of the men (probably the high-spirited Frank Demonico) showed the split-second timing of a talented wit: "Oop—Jackie Robinson just hit another home run!" On another occasion, Mont Burhans delivered a lecture on the physics of floatation and mass that was memorable. The proposition was that a fish, when placed in a bucket of water on a scale, would not affect the reading on the scale's gauge—so long as the fish did not touch the side of the bucket. In consideration for Mont's stature in the plant—of which he was also a director—as well as his age (late sixties), there were no derisive hoots of disbelief that most certainly would have been raucously emitted for anyone else positing such a proposition. Quite to the contrary, there was a sudden, stunned silence, followed by flickering sidelong looks and hurried stuffing of sandwich wrappers into lunch boxes.

The Heart of a Hudson River Brick Plant

Other indelible memories, not only for myself but also for anyone else who was ever connected with the plant, were the sounds of the brick machine and the whistle on the boiler room stack. Anyone familiar with the brick machine's rhythmic sounds would agree that it was the heart of the plant, with its "whump-clank". . ."whump-clank" at intervals of approximately two and one half seconds. Every man who ever worked at the plant probably had his own rendition of that sound. Each "whump-clank" signified the moulding of ten bricks. Another signature sound from the machine room was the loud "thwock" report of the mould bottoms being smartly struck, at the moment of mould's mechanically controlled inversion over the palette, with a long piece of wood. This indicated that the clay was sticking to the mould surfaces. Striking the mould bottom was supplementary to the regular impact of mechanical hammers (knockers) at each end of the moulds, which routinely loosened the clay.

In the machine room, a long steel stair led from the upper level, where the clay cars were dumped into the granulator trenches, down to the main brick machine level—right past three very large, howling electric motors (for driving granulators and conveyors) that were at ear level. A high school friend of the author's—a big, strong football player type—was taken to the plant for a visit, including a trip into the machine room that went down those steel stairs. Realizing that he wasn't right behind, a backward look showed him standing absolutely motionless at the top of the stairs—transfixed by the relentlessly stamping brick machine, conveyors loaded with clay and pallets, howling motors, quivering drive belts, etc.—not exactly a law office—and a sight that had been quite unimaginable to him.

The boiler room whistle, signaling the start of the day's work, noon, etc., had a sad and vaporous note due to the low-pressure boiler to which it was connected—definitely not the imperious shout of the whistle on a steam railway locomotive's high-pressure boiler. There was an old story around that the plant's whistle made some of the workers from the South so unbearably homesick that they left well before the end of the season. The boiler room was its own unique scene, quite serenely another world. The two big boilers (ten to twelve feet high) always got the highest ratings for interior cleanliness (very important for boilers) from the state boiler inspectors. Along one wall, the brightly burnished connecting arms on a pair of small condensate return pumps moved indolently back and forth while the pumps emitted gentle hissing sounds.

The sight of a kiln firing was unforgettable and particularly dramatic at night. Looking into all those arches was truly to see hell's fires and the kiln emitted its own arrhythmic, low-pitched "whoomph" sounds. Modern tunnel kilns simply don't have such dramatic visual impact—it's like buns going in and out of an oven in an automated bakery. Some of the kiln burners devised a cutting-edge school of 'kiln cuisine,' producing splendid roast turkeys and other delicacies in the arches. Collaterally, the kiln shed transfer locomotive operator, Marshall Kithcart, was not to be outdone. He resourcefully prepared such fare as lunchtime sautéed pork chops on top of the locomotive's hot engine block (the chops were wrapped in aluminum foil)—the birth of 'engine block cuisine.'

Change at Dennings Point and Jovas'

The new conditions in the Hudson River brick market also affected other manufacturers. General Builders purchased the Dennings Point Brick Works from the Ramsdell Estate in 1946, doubling its capacity with the addition of a second brick machine, a car dryer, and additional kiln sheds—all in anticipation of the Stuyvesant Town project. Similarly, the Jovas added a second tunnel kiln, thereby discontinuing use of scove kilns at that plant.

A Collegial Association of Manufacturers

During that postwar period, there was a condition of collegiality among most of the manufacturers, holding each other in considerable respect and esteem. This was well-deserved and allowed notable cooperation. Even before World War II the erection of an enormous sign with ten-foot-high letters, mounted on the Jova's kiln sheds, bearing the message **BUILD WITH BRICK**, was evidence of this. This sign was advice to the passengers on the main line of the New York Central Railroad (which had heavy passenger traffic at the time and included splendid luxury trains like the Twentieth Century Limited). All the Hudson River Brick Manufacturers Association members subscribed to that sign. Such cooperative spirit stood in sharp contrast to the edgy selfishness of the late-nineteenth and early twentieth centuries. Admittedly, the drastic reduction in the numbers of manufacturers may very well have been a factor in creating the new collegiality. Far more importantly, this spirit extended to an enlightened initiative that was notable in any circumstance.

In the late 1940s, the manufacturers shared a common interest in the use of a pulverized shale addition to the clay mixture. It was essential to obtain the best technical assistance for investigating this process and to thoroughly review other possibilities for improving the product. The competitive position of Hudson River bricks had to be secured to the fullest extent of their capabilities. In addition, there was considerable interest in upgrading production facilities—and the manufacturers wanted guidance in this matter as well. Add to that the matter of the obvious narrowing of the market to face bricks and the implications of unknown changes and competition in the future. In a move to effectively deal with these issues, the pre-World War II association with the New York State Ceramics College at Alfred University was renewed, but in a far more ambitious form. The best technical and scientific expertise available was sought, and it was done in the best way possible. Contrary to one critic's analysis—who stated that the manufacturers sought assistance when they were in direst trouble and went to academe on bended knee to seek a way out of their dilemma—this was done at a moment when the industry was at the top of its form and profitability. Such a characterization of ineffectuality could not have been more incorrectly assigned, and it is a slight to very competent and intelligent men whose actions in this matter testified to their capabilities.

The manufacturers agreed to fund a regional laboratory, complete with staff, testing facilities, etc., to begin operations in 1949. Van Hutton put all investment plans on hold pending results from the new laboratory. The laboratory was housed in a building made available, gratis, by Harry Weinberg, owner-president of the East Kingston Brick Company. Weinberg was a New Jersey accountant by trade and a building materials dealer and labor relations expert, who was also considered to be a very constructive member of the brick manufacturers' group. His acquisition of the East Kingston Brick Company was an instance of a talented outsider coming into the Hudson River brick business at a prosperous time, as was General Builders' acquisition of the Dennings Point.

High Quality Industrial Research

The laboratory director was the best obtainable, in the person of Charles Amberg, director of the laboratory at the Ceramics College of Alfred University. Testing equipment was quickly installed and staff immediately hired. The publication of monthly reports for the manufacturers began in July 1949. Initial studies centered around problems recognized as common to Hudson River bricks, most particularly in meeting certain standards of the American Society for Testing Materials (ASTM), the principal U.S. institution for the establishment of product and materials standards. (Not alone, ASTM shared this function with the U.S. Department of Commerce and the Underwriters Laboratory for fire safety and electrical equipment standards). It was necessary to deal with the standard relating to brick because government housing authorities and architects used certain criteria in creating specifications and requirements for materials used in construction.

Weathering : Truths, Half-Truths, and Materials Standards

The issue of water absorption relates to durability under numerous cycles of freezing and thawing. On the basis of water absorption, Hudson River bricks were rated MW under the ASTM standard[1]—indicating a higher rate of absorption than the SW bricks. The lack of correlation between absorption and damage by freezing, in Hudson River bricks, had been observed in the late 1930s at the Alfred University Ceramics Experiment Station. Testing conducted at the East Kingston laboratory in 1950 confirmed this—the higher rate of water absorption by Hudson River bricks, as indicated by use of the ASTM-specified tests, did not cause correspondingly greater damage from freezing. Quite to the contrary, test results showed Hudson River bricks to be surprisingly resistant to frost damage. However, the ASTM standard made higher resistance to freezing implicit in low water absorption—and vice-versa. Higher susceptibility to frost damage was implicit in high water absorption. The two separate sets of laboratory data showed that Hudson River bricks, rated MW, suffered far less damage from freezing than implied by that standard, with its corresponding implication that SW bricks were more durable. The reasons for the Hudson River bricks' durability is related to the very same large pore spaces that do indeed allow the relatively larger water absorption, thereby relegating them to MW grade. Those pore spaces are also sufficiently large to allow for rapid drainage of water out of the brick, as well as providing space for the expansion of freezing water that may remain in the pore spaces. The self-draining capabilities of those bricks were established by the 1930s Ceramic Experiment Station laboratory by observing the phenomenon in bricks stored out-of-doors for the winter, as well as in laboratory tests. Those same winter observations also revealed that the Hudson River bricks actually retained significantly less water than stiff mud shale bricks. Large pore spaces also allowed far less water penetration due to capillary action than small, fine pores—and larger pores tended to retain water for a shorter period of time. The East Kingston laboratory director, Charles Amberg, went to great lengths with the appropriate ASTM subcommittee for the purpose of modifying their standards in order to allow for structural clay products (bricks) that had proven resistance to damage from freezing, even though they did not correspond to the assumptions of those standards. A waiver has been inserted into the later editions of the same standards that permits a manufacturer to submit test results on MW bricks, showing that the tested samples were not damaged in excess of allowed limits. The qualifying bricks were considered to effectively meet the higher SW standard. In addition, both of the same ASTM standards eventually specifically allowed the use of MW bricks in those parts of the United States designated as having severe weathering conditions, where exterior brick surfaces were not in contact with the ground. Earlier versions of those same standards, in effect during the 1950s and 1960s, did not make such distinctions and included language that could be used to specifically exclude MW bricks from those parts of the country that had severe weathering conditions. Such exclusive language was removed in later ASTM editions of the 1970s—after the damage had been done to the Hudson River brick industry.

1 ASTM C62-75 for building brick and C216-77 for facing brick.

Specifically, the effects of the standards' requirements for Moderate Weathering (MW) and Severe Weathering (SW), with respect to Hudson River bricks, were the objects of the laboratory's initial efforts. The principal elements addressed by these requirements are those of water absorption and compressive strength. The Ceramics Experiment Station had examined water absorption in the late 1930s, and the New York City Bureau of Buildings noted similar testing (Chapter 3) in the early years of the century.

The Color of Brick

Already under competitive pressure from buff-burning Ohio and Pennsylvania bricks, the manufacturers also wanted the laboratory to do very early investigations of various means of effectively coloring the bricks—because customers wanted some other color than red. The use of mineral additives to the molding sand, including manganese (grey- and brown-burning), rutile (giving a buff color), aplite (giving a subdued pink), as well as white sand quickly and successfully accomplished a range of color selections. The new range of colors gave the Hudson River bricks a significant boost in the market. The colored molding sand was fused to the brick body during firing.

Investigating the Use of Shale

In addressing one of the principal purposes for the establishment of the laboratory, extensive analysis of the shale resources of the Hudson River and various means of utilization was also performed, with varying results for different manufacturers. The investigations also showed that high-quality shale was in short supply. The New York State Department of Commerce furnished a geologist for the study. The only genuinely superior shale on the river was located at Newton Hook, on the east shore, north of the city of Hudson. Implicit in the study was the correlation of the manufacturers' location with economical access to quality shale resources. To the same extent as the river's clay resources, much of the shale also contained considerable amounts of calcium—a harmful prospect to most of the manufacturers.

Laboratory tests did indicate that the addition of good quality shale materials to the clay mixture, as a sand substitute, could give significantly lower water permeability and higher compressive strength. However, a large percentage of the economically available shale had excessively high calcium content. This caused the East Kingston Laboratory to note the certainty of the problem of the short firing range, which also occurred in the production of extruded bricks made from this shale. Two manufacturers, Jova and Dennings Point, concluded that the possibility of good results justified the considerable investment required for shale grinding and extrusion equipment, allowing them to make stiff mud brick. Dennings Point used a combination of seventy-five percent shale and twenty-five percent high-grade yellow clay. Their on-site shale had an unusually long firing range. The Jova plant made an all-shale brick. Dennings Point also installed a tunnel kiln. Subsoil conditions, created in the late-nineteenth century, continually daunted the kiln. A very large amount of inferior fill had been used to create a significant increment to the area of the waterfront property (belonging to Brockway at that time of the filling operation). Because of the unstable subsoil conditions, that tunnel kiln's foundations were unable to provide sufficient support for the prevention of settlements that adversely—and seriously—affected its performance for the rest of that plant's existence (frequent kiln car derailments required the tunnel kiln's operation to shut down). The Jova plant's upgrades were eventually regarded as having provided a moderate improvement of their product—this because the economically available shale was itself of marginal quality. The very admirable desire to invest and improve didn't always have the desired outcome.

Discouraging Findings and Limited Improvements

Having eagerly anticipated guidance for a considerable program of investment for upgrading its product, the Hutton Company received disheartening news. The economically situated, nearby shale deposits, classed as 'Esopus' shale, were contaminated with Devonian limestone, laced with calcium, had poor workability (excessive silica content), and provided an unsatisfactory color when fired. Even a relatively nearby shale of somewhat higher quality (Davis shale from the Kingston West Quadrangle) did not provide a significant reduction in absorption tests that the laboratory performed. As a progressive, investment-minded manufacturer, this analysis must have been very difficult for Van Hutton to accept—and unquestionably raised very unwelcome concerns about the long-term future of the plant. (The Kingston Laboratory's less-than-helpful recommendation urged the use of shale "equal in quality to Newton Hook shale." That material was unique on the entire river and a matter of thirty miles north of Kingston.)

Several years went by before a final decision was made in 1955 to acquire some high-quality clay property only three miles north of the plant in order to augment the on-site clay at the plant. The existing clay quality had shown signs of deterioration with further excavation westward into the clay banks. The new clay was trucked down to the plant, to be added in relatively small amounts because of its very 'strong' quality. With two different qualities of clay already in the plant's banks (the 'lower bench' had a considerable content of fine 'quicksand' clay), the mixing process now had to combine three different qualities of clay. The wisdom of the installation of the two separate granulators during the 1920s allowed the strong new clay to be added to the weaker quicksand clay, in a controlled manner, in one of the two granulators. The laboratory's recommendation to the Hutton Company for greatly increased mixing of the three different clays was carried out with the 1955 construction of an addition to the machine room. This was a powerful new vertical pug mill that received the clay from the granulators, and then conveyed it to the brick machine.[4] The clay from the newly acquired property had both different thermal characteristics and different water

Figure 76. Hutton Company Clay Banks, 1950s. Electric Thew shovel from the 1920s modernization is seen at left of clay car train at lower middle of picture; these slower, older shovels were displaced by new shovels for most loading, the older types being relegated to less frequently used parts of the bank. (Photo by Hutton Company)

retention properties from the on-site clay, making thorough mixing a particularly urgent matter. The J. C. Steele Company of Statesville, North Carolina, manufactured that pug mill. J. C. Steele Company was a fateful name in the immediate future history of the Hudson River brick industry and will reappear in the following chapter. Although without the beneficial effects of suitable shale, this new mixing capability—combined with the newly acquired high quality clay—provided a notably successful improvement in quality, as attested by several entries in directors' minutes. The plant continued to make a soft mud brick of clay and sand.

Fine Sand—Excavated by Hand

Regarding clay bank management at the Hutton Company, fine sand for brick mold dusting occurred in identifiable veins of such minimum width that the fine molding sand could not be removed with power equipment without the introduction of coarser adjacent material. This fine sand therefore had to be shoveled by hand into the clay bank rail cars right up to the end of the plant's operation in 1980.[5]

The Intractable Problem of Materials Standards

Three years after commencement of the East Kingston Laboratory's operations, the importance of the use of high quality shale for the reduction of water permeability became more urgent with the first strong criticism of Hudson River bricks by the federal and state housing bureaucracies. That criticism concerned that part of their materials standards for the types of projects which had virtually become the life blood of the industry. Their standards incorporated the ASTM standards for both compressive strength and permeability. For compressive strength, the difference between the MW grade (including most Hudson River bricks) and the more stringent SW grade was the difference between being able to build a wall one half mile high (using the MW standard) and building a wall that could be fourteen percent higher (using the SW standard), before compressive failure of the bricks at the bottom of the wall. The compressive strength standard was of less critical significance—and Hutton Company bricks met the SW compressive strength standard anyway.

The permeability standard was the greatest worry for most of the Hudson River industry. The porosity of the bricks prevented the industry from attaining the SW rating. The most principal reason for the inability to attain lower water absorption was the already much-discussed calcium contamination of the clay, as well as much of the shale resources. The resources did not allow use of the higher kiln temperatures, which could obtain sufficient partial fusion of the clay particles, thus reducing the volume of the voids (pore spaces) between those particles. The day of reckoning for the Hudson River brick industry, on account of the calcium contamination, was rapidly approaching. However illogical, the limitation of the use of bricks to SW grade for government subsidized housing gradually increased over the next decade. Specification writers for nongovernment work tended to follow suit.

The Possibility of a Radically New Brick

One sympathetic critic, William Minnock, raised a reasonable point of conjecture about an opportunity that the Hudson River brick industry may have been underrated, namely the manufacture of the so-called "SCR" brick (5-1/2 inches wide by 9-1/2 inches long by 2-5/8 inches high) that had provisions for vertical reinforcing steel running through the height of the wall. Most building codes allowed this brick for single-story bearing wall construction of a single thickness of bricks (the bricks supported the roof loads) that was useful for light industrial types of buildings, single-story housing, etc. It was definitely known to the local industry—the author remembers GVDH's positive and interested mention of it—but only Powell and Minnock produced it. One reason may have been the uncertainty of an adequate market for such a relatively radical product that had been principally used in the South. However, Powell and Minnock did manufacture this type of brick for two years, finding an active single-story housing market for them in Boston and on Long Island.

Labor Relations

Post-World War II labor relations generally paralleled those in other industries, including the arrival of paid vacations, time and a half for work in excess of forty hours and, eventually, family medical benefits. There were only two insurance companies that were willing to deal with coverage for brick industry personnel. One of those insurers added a surcharge for non-whites who were on the payroll. Aetna, making no such differentiation, became the insurer of choice.

Union success in obtaining higher wages for Hudson River brick workers also put the river's industry at a significant cost disadvantage with respect to other union jurisdictions in New England, New Jersey, and Pennsylvania, all of which the United Clay Workers Union represented. The latter union was also notable for having been considerably less confrontational than the local ILA Brickhandlers Union. Such posture had negative consequences for the Hudson River industry. Circumstances eventually changed that confrontational attitude—which was inappropriate considering a large majority of the manufacturers were very fair-minded men. Generally, the relationship with the Union of Operating Engineers was more cooperative and their leadership became far more sensitive to the manufacturers' cost circumstances during the 1960s.

With a change of leadership in the 1960s, the Brickhandlers Union followed suit. That union initiated strikes at the plants—those with the worst labor relations going out first. The Hutton Company was always among the last to go on strike. Blandishments to workers from plants not yet on strike included an appeal to their common Italian ethnicity, an approach that fell on deaf ears at the Jova plant with its Hungarian and Irish workforce. One tactic that the unions conceived for contract negotiation sessions was to run all-night meetings in a smoke-filled room in New York City, in an effort to simply physically wear down the manufacturers, all of whom were well into middle age. Labor's representatives were much younger men, and the Operating Engineers' Pete Weber looked like a World Wrestling Federation performer (not to diminish his skill and fairness as a negotiator and his own brand of amiable charm). The manufacturers did effect a change in the locus of the meetings to a big function room in the Governor Clinton Hotel (big enough so they couldn't get smoked out) in Kingston, where they also rented a room that allowed for a quick nap during those all-night sessions.

Importantly, large increases in productivity, throughout the U.S. economy, allowed noninflationary price increases on manufactured products that offset the labor cost increases—and this applied to the Hudson River brick industry for a decade after World War II. That pricing flexibility began to vanish after the mid-1950s. The subject of labor relations raises the instance of the final moment of the existence of the industry's 'bad boy' plant (which shall remain unnamed). The owner assembled the workers and informed them that unless they took a pay reduction, he was going to close the plant. One of the men inquired as to what he, the owner, was willing to sacrifice. After careful thought, the owner replied that he would give up his chauffeur and drive his own car. The men walked and the plant shut down—permanently.

Significant Productivity Increases

A final word must be said about the productivity increase that the Hutton Company accomplished due to improvements to the plant, essentially an example of 1920s modernization. One scholarly critic claimed that the brick industry's productivity increase did not justify the cost of investments incurred from modernization.[6] In response to this, recognition must be made of the late-nineteenth century standard of production at an average of one thousand bricks per day per man during a workday that averaged twelve hours.[7] In the post-World War II period, productivity rose to approximately nineteen hundred bricks per man per day at the Hutton Company—an increase of nearly one hundred percent during an eight-hour workday. Adjusting for the greater length of the nineteenth-century workday (twelve hours average per U.S. Census *Products of Industry*), the productivity increase is an improvement of almost one hundred ninety percent, equivalent to one man in the modernized plant doing the work of almost three men in the pre-modern yard.[8]

One Plant's Changes of Generations

In the Hutton Company family, William Tierney, the plant superintendent who had joined the company in 1906, retired in 1953. One critical employee judged him as being "good with the men"—meaning that he was skilled at obtaining willing cooperation. Tierney's replacement was Tom Turck, the brick machine foreman whose father, Frank, had joined the company in 1904. Tom, a driven perfectionist, had total knowledge of the plant's operations—and most particular knowledge of the heart of the plant—its brick machine room. That big, single brick machine was literally the heart of the plant, and its operation had to be absolutely reliable and uninterrupted. Without it, there would be no bricks. Turck's skill and meticulous drive assured the machine's very reliable operation for a considerable number of years. As previously noted, Tom also became a director of the company. The year 1953 saw the institution of profit sharing for supervisory personnel. In the following year, a formal ceremony, including the presentation of watches, honored Sam Frangello and Frank Turck for fifty years of service to the company. Such long-time associations, dating from Van Hutton's boyhood, were highly valued. Prior to 1956, directors' minutes had made many references to the notable efforts of 'key men'; in that year, the first entry referred to the notable efforts of all employees. On numerous occasions, GVDH related to the author (working in California at the time) the many accounts that showed his great involvement in that ongoing, highly constructive dynamic with the men, such as, "John Polacco has been coming up with some really good ideas—quite good thinking!" GVHD's relationship with the men recognized that this was a tough business—very similar to building construction in its physical demands—and that business did place its demands on everyone engaged in it. It was hard work, in which men were mutually reliant. One man's tight-lipped estimate of GVDH: "He's a man of his word."

Directors' minutes of 1954 made note of good prospects for the business, and that was the last time such an entry was be made.

In one of fate's gestures of prescience, the end of the prosperous post-World War II decade (1945-1955) occurred simultaneously with the death of Van Hutton's memorable wife of thirty years. She died of a sudden heart attack at age fifty-seven. The optimism of the previous decade suffered a critical blow, and the events that followed confirmed very different conditions for the Hudson River brick business.

Notes

1 Figure 71. Priester fork and dryer car.
2 Figures 59 and 60.
3 Wayne Brownelle. *Clays and Shales of New York State*, 1950.
4 Hudson Valley Brick Manufacturers Assn. East Kingston Laboratory. *Special Report for The Hutton Co.*
5 Francis Duffy. (See Bibliography).
6 See Appendix – PRODUCTIVITY
7 A latter-day writer states that "a worker was expected to produce about 1,000 molded bricks per day. . . ." implying that each worker had a piece rate for personally hand-molding that number of bricks. That rate is, in fact, derived arithmetically: (total production of bricks) divided by (total number of employees, no matter what their tasks). Examination of production statistics in the 1880 U. S. Census *Products of Industry* shows evidence that the generally assumed 1,000 bricks per man per day average was on the high side, with 800 bricks being a common figure. At any rate, this is a general ratio that relates total manpower required to manufacture a specific measure of production (1,000 bricks). Mechanization and automation reduces the number of workers, thereby increasing the output per worker.
8 For eight-hour day: 113,500 bricks divided by 60 men = 1897 bricks/man/day; Adjusting for a twelve-hour day: 1897 divided by 0.66 = 2,866 bricks/man/day.

Principal Sources for Chapter 5

Amberg, C. R. *The Development of Soft Mud Brick In The Hudson Valley Region.*
American Society For Testing Materials (ASTM). *Standard Specifications for Building Brick,* C62-58 and C62-75a; *Standard Specification—Facing Brick,* C-216-60 and C216-77.
Brownell, Wayne. *Clays and Shales of New York State.*
Duffy, Francis. (See Bibliography).
Haskins and Sells, CPA. *The Hutton Co.—Condensed Report on Examination For the Year.* 1953.
Hudson River Brick Manufacturers Assn. Reports of the East Kingston Laboratory, 1949-1952.
Hutton Co. Minutes of the Stockholders and Board of Directors.
Hutton, GVD. (Statements made to the author).
McMahon, Kolstad & Barnes. *A Critical Study of the Manufacturing of Soft Mud Building Brick.*
Minnock, William (See Bibliography).
Polacco, John. (See Bibliography).
Siebert, Robert. *Dennings Point Brick Works, 1946-1968.*

CHAPTER 6
Too Many Problems At Once: The Death of Three and a Half Centuries of Brickmaking—1957-2002

THE BEGINNING of this final period of the brick industry, in 1957, includes an event that harked back for fifty to sixty years. Accountants' annual reports for the Hutton Company for that year—and for 1958—showed a drastic drop in sales, with negligible profit (a slight loss in 1958). Note should be made of serious labor disruptions in the New York construction industry at that time. Others in the brickmaking community must have had a similar experience, insofar as four of the manufacturers (Jova, Sutton and Suderley, Powell and Minnock, and Hutton) pursued the possibility of consolidation. This elusive strategy that was always beyond the reach of the river's brickmakers, in this instance used the services of Kidder, Peabody (Louis Thomy as leader for the study). The proposition was to acquire an existing plant in the vicinity of Perth Amboy, New Jersey, a center for face brick manufacturing. This would have provided uncontaminated clay and good access to known markets. Serious study was made of this possibility. A directors' minutes entry eventually was made that, "related the disappointing nature of Mr. Thomy's findings," and nothing more came of this matter. One account states that the sellers of the New Jersey plant were using the Hudson River manufacturers' interest as a means for obtaining a higher price from another buyer. At any rate, leaving the Hudson River was difficult for those men to contemplate.

Some Old Companies Fall Away

During the 1950s, longstanding brickmakers disappeared forever, including Terry (founded 1850), Brigham, Staples, and Sutton and Suderley (bought out by Powell and Minnock) – all originating in the nineteenth and very early twentieth centuries. The last president of the Staples plant, in Malden, reappeared as the final owner of the Hutton plant.

Invaluable Help in this Writing

At this point, acknowledgement must be made to William Minnock, former president of Powell and Minnock, now retired (see Bibliography), for the indispensable contribution to much of the balance of this chapter. This gentleman's tenure in the brick business overlapped with steam-driven equipment (immediately after World War II), extended through many years of soft mud molded brickmaking with the use of scove kilns, and included a total transformation of that plant to an all-shale, extruded brick plant with two tunnel kilns. In addition, his background was greatly expanded after the sale of the plant to General Dynamics (for a subsidiary of which he became a vice president), thus giving him significantly broadened perspectives of the brick industry outside the Hudson River region. Also integral to this account is the tape recorded testimony of Ten Eyck Powell (recorded in 1990), thereby giving many valuable insights on this period. That gentleman was a partner and vice president in the firm of Powell and Minnock.

Rapidly Declining Profitability

Hutton Company records show a big recovery in sales volume in 1959—but there was a major flaw. Profitability of the business had been reduced by a third, and that was only the beginning of the rapid erosion of profits. Directors' minutes state above-average operations relative to other Hudson River plants, tight profit margins while operating at capacity and market conditions that precluded price increases. During the years 1960 to 1963, there were periods of time when the plant operated for six days a week with limited overtime. Directors' minutes were also making regular note of the high level of workforce performance, which was now essential to the plant's existence. There was no lack of a tenacious willingness to tough it out on the part of all hands.

The increasing reduction of profit margins was principally a function of the new order of construction site practices. For a very long period of time, common

brick had literally been dumped at the construction job sites, and laborers sorted out the different grades for use in various parts of the walls. The contractors assumed the labor costs. The dumping also damaged some number of the bricks. With the arrival of the exclusive use of face brick after World War II, the dumping process continued. The contractor again assumed the cost of assembling bricks into manageable quantities for various locations on the job (scaffolds, etc.) and for removal of damaged bricks.

After the mid-1950s, job-site demands drastically changed, and brick manufacturers had to bear the consequent costs. Henceforth, bricks were required to be delivered to the jobs in 'packages' of various quantities. They were bound together with steel straps that an ingenious, portable machine, having its origins in the 1920s, tensioned in place. The prime maker of such equipment, the Signode Corporation, advertised its strapping equipment in brick industry trade journals at that time for use in assembling small numbers of bricks for delivery to architects and contractors as samples. The technology then greatly improved, in order to allow secure binding of the much larger quantities that became commonplace by the end of the 1950s. Signode is still a leader in the field. The strapping machine is now made light enough to be manually placed directly over each strap. The mechanism tightens the strap, securing the overlapped ends with a light metal pressure fastening and cutting the end of the strap snug against it. Needless to say, that small machine exerts considerable force in tightening the straps and in securing the fastening. The various quantities of bricks for different job conditions must be hand-placed on wood pallets of various designs before strapping.

'Superlight' packages were for residential building supply yards; high-rise scaffold packages became standardized, and other packaging requirements were met as well. The typical package was 530 bricks. Bear in mind that packaging commenced at the end of the firing process, and the bricks were available to the packagers at tunnel kiln and scove kiln plants in two different configurations. At the tunnel kiln, small units of bricks on the tunnel kiln cars, requiring minimal movement by the worker for removal from those cars, were assembled into packaged units, with minimum bending and movement for the worker. Much more laborious was assembly into packages at the scove kiln face where the bricks were distributed over the surrounding immediate area, requiring much bending and movement for the worker. At the Hutton Company and its successors, bricks were assembled on a pallet and then taken by a forklift to a packaging area—all far more laborious than the simple arrangement at the tunnel kiln plants. However, the Hutton Company procedure made the best of an adverse situation.[1]

Figure 77, facing page. Kiln (at former Hutton Company), late 1970s. In process of being taken down for shipment; strapping machine is visible on top of unit at bottom middle of picture; vapor visible in roof vents is caused by 'water smoking' (an initial phase of firing) from kiln in adjacent bay of shed; 'double coal' bricks that were placed at exterior surface of dried bricks are visibly projecting past fired units at right kiln wall—they are lighter color. (Photo by Tripoli)

With its far more efficient delivery of bricks to packagers, tunnel kiln packaging costs were fifty percent lower than costs at scove kilns. The raw numbers are inexorable—five to six thousand bricks packaged per man per day at the scove kilns versus fifteen thousand per man per day at the tunnel kilns. William Minnock was able to make the most immediate kind of comparison of these circumstances, insofar as scove kiln operations overlapped with the new tunnel kiln operations at Powell and Minnock for a decade—and he states that packaging costs were disastrous, beyond any question, for the manufacturers with scove kilns. Minnock also stated that high packaging costs were the cause of the 1973 cessation of operation of the Powell and Minnock scove kiln. The scove kiln (and one molded-brick machine) persisted in operation until that time—for almost a decade after the installation of their first tunnel kiln. This was a testimony to the effectiveness of the scove kiln—and one more final answer to the criticism of Heinrich Ries in 1900! Not being foolish romantics, Powell and Minnock would not operate a scove kiln just for old time's sake—it had to be a sufficiently paying proposition. Most critically, there was no market tolerance for the raising of brick prices in order to offset the big cost increase imposed by mandatory packaging. The rapid narrowing of profit margins inhibited investment and further damaged prospects for the Hudson River industry.

Powell and Minnock Beat the Odds

At this point, it is timely to consider the intriguing instance of the one manufacturer that could cross fate's path and successfully do what was forbidden to the other manufacturers. In 1963, Powell and Minnock of Coeymans made the large investments required for totally rebuilding their plant, based upon the discovery of a shale deposit (geologically classed as Mount Marion shale) six miles west of their plant (at Coeymans Hollow). This shale had a sufficiently low calcium content to permit thorough firing at temperatures that allowed the fusion of clay particles—no problem whatsoever for meeting and exceeding the ASTM–SW standard (Coeymans clay had the highest calcium content of all the plant sites on the river). Without economically available shale of this quality, such investment would have been unthinkable. Two to four big trucks were able to provide sufficient material to the plant.

This was to be an all-shale brick, formed by extrusion and fired in a tunnel kiln. The highly regarded English firm of Swindell-Dressler designed the entire plant. The latter of those hyphenated names is familiar from the 1920s as an early innovator-manufacturer of tunnel kilns. It was Phillip Dressler who enabled a quantum increase in tunnel kiln performance through his introduction of circulating blowers in those kilns after World War II. The extrusion machinery for Powell and Minnock was furnished by the J. C. Steele Company of North Carolina (already cited in Chapter 5 regarding pug mill machinery for the Hutton Company). Swindell-Dressler designed the entire plant and made all equipment selections using J. C. Steele's equipment on account of its peerless reputation. The dryers and the tunnel kiln were furnished by Swindell-Dressler. Phillip Dressler took a personal interest in the Powell and Minnock installation because of the high silica (sand) content of the Coeymans Hollow shale—unique within his extensive experience. The silica presented a challenge, not only because of its aggravated wear and tear on the machinery, but also because of its sudden expansion at 1,060 degrees Fahrenheit, which required a very controlled temperature increase through that level. This also introduces two more personages into the story. Richard West, a member of the ceramics engineering faculty of the Ceramics College at Alfred University, was an accomplished ceramic scientist who Swindell-Dressler retained for opinions relating to Hudson River conditions. West was familiar with this subject. He was also a leading tunnel-kiln expert, eventually performing all the clay analyses for Swindell-Dressler's worldwide operations. The English company, with a considerable history of worldwide operations, came away from that job with the opinion that West was among the very best of the consultants with whom they'd ever worked—a top man who was very pragmatic and resourceful in his advice, and very personable as well. In a tribute to Hudson River brickmakers, Richard West told William Minnock that the river's brick manufacturers had gotten as much out of the local clay as was humanly possible.

The other person to coincidentally appear was Francis Pixley, originally of Newburgh, a cousin of the Jovas and an Alfred Ceramics College graduate. Pixley, operating from Colorado, developed a very successful business as a representative for firms in the clay machinery business. Swindell-Dressler was a principal client at that time. Pixley's wide-ranging business required the use of a personal aircraft, resulting in a fatal crash that was recent to this writing.

Powell and Minnock were thereafter positioned to manufacture a quality of brick that was precluded to any other brickmaker on the river—and they were no longer manufacturing the classical molded, scove kiln fired Hudson River clay brick (with exception noted above) that had dominated Hudson River brickmaking. Their raw materials strategy bore striking similarity to that of the early-twentieth-century Kaaterskill Paving Brick Company of Catskill (discussed in Chapter 3), with its rail transportation of a similar type of shale (classed as Hamilton) from Cairo (nine miles west of Catskill).

Powell and Minnock also had been developing some other strong suits, in offsetting some of the growing New York City market problems of the late 1950s, with their entry by truck shipments into the more profitable New England market (thirty to forty percent of their production). Truck shipments to that market were facilitated by the conveniently situated New York State Thruway entrance at Selkirk (three miles distant by a good, under-populated secondary road). This was an option that was virtually impossible for the Hutton Company because of its remote location from the Thruway. The same means was used for developing markets in Syracuse and Buffalo. Truck shipments eventually became the principal method for moving their bricks to market. Powell and Minnock also trucked their bricks to a Conrail siding at Ravena for forklift transfer to railcars.

The Powell and Minnock plant was purchased in 1969 by the high-tech giant General Dynamics (maker of fighter planes, warships, nuclear submarines, etc.). The unlikely presence of this company in the brick business requires explanation, although they have ventured into quarter horse breeding as well. In the late 1950s, General Dynamics had become associated with the very successful Chicago construction materials businessman Henry Crown, through absorption of his own corporation, Materials Service Corporation (dominant in concrete materials and gypsum products in the Chicago region and still very much in existence). Henry Crown had already acquired a considerable financial interest in General Dynamics, which continues to this day (three members of the Crown family are presently on the General Dynamics board of directors). General Dynamics embarked on a campaign of acquisitions in the brickmaking field, buying up not only Powell and Minnock but also such very large entities as the Darlington Brickworks in Pennsylvania. The Materials Service Corporation division of General Dynamics, of which Henry Crown remained president, consummated the acquisition. William Minnock was confirmed as a vice president in the brick manufacturing division of Materials Service Corporation and eventually retired in 1990. In 1973, a second tunnel kiln was be added, also built by Swindell-Dressler. The Powell and Minnock plant—it continued operating under that name—was in business for the long run.

The Return of a Brawling Brick Market

Beginning in 1964, the New York brick market became a free-for-all, and out-of-state manufacturers became increasingly intrusive. Aggressive new southern competition caused Pennsylvania's very big Glen Gery Company to experience severe incursions into its traditional Mid-Atlantic markets. This same competition soon showed up in the New York market. Glen Gery greatly expanded their sales efforts in the New York market, in close coordination with the big Ohio firm of Belden-Starke (probably the best maker of buff-colored brick in the United States), as well as developing contacts with the housing bureaucracies, which included the promotion of the limitation of government agency brick specifications to ASTM Grade SW. Bear in mind that Pennsylvania and Ohio clays had none of the higher calcium content that plagued the Hudson River deposits. Equally active in moving into the New York market was the big, modern Hanley Brick Company, also of Pennsylvania. This firm also developed a similar interest in promoting SW bricks. The Binghamton Brick Company of Binghamton, New York, also promoted the exclusive use of SW bricks with the housing bureaucracies as well.

The success of that proselytizing for limitation of big government projects to SW grade became more fully realized in 1964. The New York State government became more harshly critical of the Hudson River molded bricks, showing none of the solicitude for the local clay industry that at least one of the southern state governments showed for its state's manufacturers. GVDH was directed to a mid-level Housing Authority bureaucrat to plead the case, to no avail whatsoever. After all, New York was still 'The Empire State,' and had the computer giant IBM—virtually without peer in the computer business at that time—as one of its principal industrial standard bearers. Bricks were a very secondary or tertiary consideration. The South had long since raided the northeastern textile industries and it now did the same for brick production.

One Manufacturer's Predicament

Van Hutton was quite able to see that the brick business had come to another fateful crossroads that presented the same imperatives as the 1920s—make substantial investments in modernization, or perish. His own estimate of 1963 was for a five-million-dollar makeover (year 2000 dollars) that was similar in extent to Powell and Minnock. The experience of the latter showed a cost that was close to fifty percent higher. At the location of the Hutton Company, with the shortcomings of its raw materials plainly understood, there was no possible rationalization for such an investment—and Van Hutton was in his mid-sixties. Regarding any kind of succession within the family, lifelong wariness about the long-term viability of the Hudson River brick industry had created a clear understanding that his two sons were expected to find other occupations. He recalled the upbeat words of the immediate post-World War II years, spoken to Floyd Flint, when he said to the author, "We're not going to beat the devil like we did after the war"—and that was an obviously painful prospect. Concern for the men who had so capably and unswervingly served the company became obsessive.

To GVDH, the appearance of consequential red ink was the business-world equivalent of a death warrant for an enterprise in which he had invested his talent and great commitment for over forty years. That dreaded event did take place in 1964 after several years of operating at capacity and rapidly thinning profit margins—with a forty percent drop in production from the previous year. It must be stated that the previous year's production (1963) was almost twenty percent above average, thereby making the 1964 net decrease more equivalent to a drop of twenty percent below average production levels. It goes without saying that reduction in demand forced an equivalent drop in production. All of this was enormously difficult to take for a man who had always been investment-minded and so sure of his ability to bull his way through problems by sheer will power and mental exertion—proven performance for forty-five years that included surviving the Great Depression. The year-end accountant's financial report wasn't needed to give the bad news of a substantial operating loss that had been building during the year—and the characterological impact was very evident. Such events can certainly have severe psychosomatic effects—and Van Hutton died in February 1965, at age sixty-seven, with his boots on, as they said in the old West. As one labor leader told the author shortly after the death of Van Hutton, "He enjoyed the struggle"—a fine epitaph for any man. Ironically, 1964 was a bad year for the rest of the industry as well.

The End of One Hundred Years of Family Ownership

Unknown to GVDH, because the offer was received only several days after the first of a series of strokes that would kill him within a month, the Jova Company made the first contact with the Hutton Company regarding its purchase. The sale was concluded within two months, after negotiations led by Floyd Flint (on behalf of the Hutton Company) and the Jova Company's Henry Abreu (courtly, with a lifetime of recognized competence as a hands-on plant manager). The plant's brand and name was changed to identify with the new owners (JMC for Jova Manufacturing Company). The HUTTON brand was retained for a premium line of bricks.

That transaction brought to a close exactly one hundred years of Hutton brick manufacturing on the Hudson River—1865 to 1965—the longest run of any Hudson River brickmaker under one continuous ownership. Operation under two successor owners (for fifteen years) allowed that plant to have the longest run of any plant on the river. Van Hutton's forty-five years of service to the company spanned almost one half of that century.

Peers in the brick business noted the plant's remarkable ability to efficiently produce a quality product with an obsolescent plant. One critically informed opinion stated that it had been the most efficient soft mud plant on the river. Significantly, it was envied for the quality of its workforce, and those men had such a highly-developed understanding of brickmaking—as well as stubborn commitment—that the plant could have run on auto-pilot for a considerable period of time. To some degree, two successive sales refuted GVDH's darkly pessimistic outlook for the plant's future. Those successor owners received an all-business operation that was very good at brickmaking. Perhaps GVDH didn't comprehend the persistence of the momentum that he and the men had together created.

Southern Brickmakers Invade the Yankee Market

The penetration of southern bricks into the New York market is a story in itself—and it was the final big event that brought down the remains of the Hudson River brick industry. It was not just the straw that broke the camel's back—it was an iron bar that was brought down with great force. Ironically, the northern manufacturers joined with their southern counterparts in an ambitious postwar campaign to promote the use of brick masonry, some years before the troubled times ahead. Centered in North Carolina, a concerted drive to expand brick manufacturing and market share began in the 1940s and built momentum through the next decade. North Carolina manufacturers had several very big advantages; the foremost was large amounts of absolutely premium quality clay. Their material was derived from the second principal means of clay formation, giving it the geological classification of 'residual' clay.

Northeastern clays, formed through sedimentary processes, which included glacially-derived materials such as limestone flour and transportation by water over some distance from its origins as weathered igneous rock, were anything but homogeneous. In North Carolina, clay beds, varying from 80 to 140 feet in depth, directly overlie the igneous (feldspar) rock from which they are derived. Weathering created these clay beds—no stream transportation with other glacial deposits added along the way, etc. Calcium score: North Carolina (average) 0.6 percent; Hudson River (average) 4.9 percent and as high as almost 8 percent. The Tarheels' slogan: "It is difficult to scoop up a bucketful of North Carolina clay that will not result in first quality face brick." Furthermore, this clay has a larger particle size, making it ideal for the efficiencies of extrusion, and it needed none of the sand admixture required for Hudson River clay drying because of its small particle size. Nor did the Carolina material have the high silica content of the Hudson River shales, thereby causing far less wear on machine parts. This low silica content also allowed a rapid kiln temperature increase (to vitrification temperatures) that resulted in a twenty percent reduction of firing time (relative to Hudson River clay and sand mixtures), thereby consuming less fuel—another cost advantage. Additionally, their clay provided little resistance in the extrusion machinery. It is important to remember that this was not shale, which required an extra step of pulverizing in order to render it workable. Recall that Hudson River clay structure did not allow extrusion.

The J. C. Steele Company (Statesville, North Carolina) was an important contributor to development of the North Carolina industry. It was a high-quality brick machinery manufacturer in that state since the end of the nineteenth century (its advertisements prominently featured in *Brick* during the first decade of the twentieth century). That company had a considerable interest in the expansion of the local industry in order to showcase its equipment—and that was, most ironically, the same company that furnished the big pug mill to the Hutton Company in 1955. Additionally, the North Carolina state government—very particularly including the governor's office—took a very active interest in the rapidly growing brick industry in their state, offering tax incentives for relocation to such prospects as Powell and Minnock when they were in the process of considering a new plant in the early 1960s.

Well-established, family-owned plants in that state offered particularly attractive opportunities to potential investors. Some of these plants had come upon severe internal dissentions within family ownerships regarding the use of profits, thereby putting the plants on the block for sale. Additionally, North Carolina brickmakers had the very great advantage of using nonunion labor, with significantly lower wages and benefits—and they also made extensive and advantageous use of piecework as a basis for wages. The list continues, and brick manufacturers in North Carolina had a record of being not only highly cooperative with each other in dealing with external problems, but also with other major southern manufacturers, such as Bickerstaff of Alabama and General Shale Products Corporation of Tennessee. This ability to act in concert allowed southern brickmakers to obtain the final cost advantage that they needed in order to break the long-standing isolation of the New York market from large-scale competition.

The Hudson River manufacturers had possessed a great advantage for a long period of time, namely cheap bulk transportation by water. This final stroke was the highly successful negotiation of very low rail shipping rates with the southern railroads, allowing large-scale penetration not only of the New York market, but also of Boston and Chicago markets.[2] As a result, a type of southern brick poured into New York that could not possibly have been made on the river. All southern brick were extruded, wire cut and with holes cored in them for lightness—common enough in the extruded brick world. The most unpleasant surprise was an 11-inches-long-by-three-and-a-half-inches-high brick that drastically reduced a masons' time required to lay up an equivalent wall area with conventional bricks (having a wall area than is almost double that of a conventional brick). Hudson River molded bricks, with the much higher moisture content required for molding, could not possibly be dried with such a length dimension without shrinkage cracking, which would render the bricks useless. Those oversize southern bricks were an instantaneous success in the New York market, and they very easily met all the ASTM specification for SW bricks, etc. The Hudson River brick industry's principal market for well over a century—New York City—was a shambles. The South won this one, as some kind of one hundredth anniversary of the end of the Civil War.

Contract Turmoil and Union Reorganization

Simultaneously, a drama was playing out that included not only one of the largest housing projects ever constructed—Co-Op City—but also two of the remaining Hudson River brick plants. Added to this was a full-scale makeover of union representation for brick plant workers. The International Longshoremen's Association (ILA) was unable to effectively continue as the parent of the Brickhandlers Union—it was in a financially precarious condition—and the workers were very mindful of a lack of effective representation. In 1966, the Brickhandlers Union turned to the Teamsters for affiliation. Tony Alecca led their Local #445. He represented a range of trades and professions (truck drivers, nurses, police, warehouse workers, etc.) in seven counties (including Westchester, Dutchess, Ulster, Putnam, Orange, and Sullivan Counties). The manufacturers fully subscribed to this new order with letters of intent from each company.

The Union of Operating Engineers, led by Pete Weber and Henry McArdle, had already provided exemplary willingness to cooperate with the manufacturers in matters of balancing the union's interests with the financial welfare of the industry and its wage costs. Alecca provided not only a willingness to understand the cost imperatives of the brick business in dealing with the manufacturers, but also provided enlightened statesmanship in his willingness to assist in preserving brick manufacturing jobs by promoting the use of Hudson River bricks with the government agencies. This was demonstrated at the beginning of the construction of the Ulster County Community College, which initially used southern brick. Alecca organized a picket line at the Stone Ridge campus site, and urged county legislators to change to a local source of bricks. That contract went to Powell and Minnock, who made a brown brick by means of the addition of manganese to their ground shale.

Mindful of the cooperative intentions of the local union representatives, it must also be stated that wages in the jurisdiction of another Mid-Atlantic union, the United Clay Workers, were fifteen to twenty percent lower than the Hudson River wage schedules. William Minnock noted this during his tenure as vice president at both the Darlington works (Pennsylvania) and at Powell and Minnock. The southern plants, with their nonunion labor force and extensive use of piecework, had substantially lower labor costs.

Subsequently, the provision of bricks for the enormous Co-Op City job became a troubled experience in fending off out-of-state brick manufacturers. The Dennings Point plant had been purchased from General Builders. Thereafter this plant was considered worthy of special consideration for the provision of brick to the Co-Op City job—to the extent of the negotiation with the union of a five percent reduction in its labor costs, a reduction that was insufficient to reduce that plant's costs to a level that would allow its continued operation. With the additional impediment of its faulty tunnel kiln, the Dennings Point plant was unable to deliver on its contract and subsequently went into bankruptcy. Thus ended ninety years of operation of the Dennings Point Brick Works, and Powell and Minnock successfully assumed the Co-Op City contract. They were the sole supplier, furnishing a 3-1/2 inches high by 8 inches long 'Econo' brick. This was a brick size well outside of any previous Hudson River brickmaking experience. The Jumbo brick only had a height of 2-5/8 inches (with the traditional standard brick height at 2-1/4 inches). The Econo brick also was a brown brick that had a manganese additive for coloring.

The Collapse of Northeastern Brickmaking

The entire universe of northeastern brickmaking and marketing was rapidly disintegrating. The entire system of waterfront brick dealers that had provided the essential link to the metropolitan market for the Hudson River brick industry was in a state of collapse. By the mid-1970s, the Hudson River brick industry was reduced to two plants—Powell and Minnock and the former Hutton Company, operated by a descendant of the same Staples family that had been in the brick business since the nineteenth century. Terry Staples, whose Malden brickmaking had terminated operations in 1958, had purchased the plant from the Jova company in 1970, seemingly as more of a sentimental *beau geste* than as a hard-headed business decision. That continued operation did preserve men's jobs for another decade. With a sudden surge in popularity of traditional molded bricks for residential use, that plant developed a surprising market in Detroit (through the Clifford Brick Company), with shipments by rail via the adjacent Hudson Cement Company's rail spur (out of service in the early 1980s) that was connected to the West Shore-Conrail track. The plant's profitability was marginal (negative, in the final year of operation) until its life ended in a 1979 by an edict of the New York State Department of Environmental Conservation that required the ending of air pollution by the scove kilns. Insofar as that could only have been accomplished with the building of an expensive tunnel kiln—with Terry Staples in his seventies—that was out of the question. So ended the life of the longest-operating plant—for a total of one hundred and fifteen years in three and one half centuries of Hudson River brickmaking. It was the next-to-last plant to go out of business in that very long period of time. The New England and New Jersey brick industries had virtually disappeared. The Binghamton Brick Company—a willing cooperator with out-of-state manufacturers who promoted the exclusive use of SW bricks with the state housing bureaucracies—was also out of business.

The Workings of Fate

Chapter 2 recounts Sylvester's 1880 history of Ulster County, with its extensive description of the Cordts and Hutton brickyard, referring to Superintendent Charles Van Wert. This writing also stated that, regarding Van Wert, there was an astonishing example of the workings of fate a century later. Present at the closing of that same plant, in 1980, was Roger Swart, a lineal descendant of Charles Van Wert. Earlier genealogy of the Swart family also includes one of the captors of the Revolutionary War spy, Major Andre. Van Hutton hired Roger Swart as a young man in 1960 as an office assistant. Roger was well-regarded and had a memory for good stories that he liked to tell about his early days at the plant. He related to the author that Van Hutton ran all of the directors' and stockholders' meeting details with the same punctilious care as a big, publicly-held corporation. Roger felt that he had gotten a superior education in the English syntax by typing of the president's letters, as well as absolutely fundamental instruction in sound business practice: "Roger, pinch the pennies, and the dollars will take care of themselves." That came from the same man who financed capital expenditures solely with retained earnings and, during the 1920s, never bought stock on margin. Roger also related the manner in which seasoned representatives from large supply houses emerged from the president's office after negotiating some sales, with a sudden realization that they had been so waltzed around that they couldn't remember just what it was that they'd agreed to. Van Hutton had a mischievous sense of humor.

With the closing of the plant in 1980, Roger looked for a position at the Powell and Minnock plant. The question put to him by his prospective employer was: "Were you at the Kingston plant when Van Hutton was President ?" With an affirmative answer, the job was his.

That top of the line brick machine, purchased from the Lancaster Iron Works in 1928, was sold to the Powell and Minnock plant and was moved to Virginia by a subsequent owner. The most recent operator, the firm of Isiklar Holding, headquartered in Istanbul, was typical of a large influx of foreign ownership into American brick manufacturing. That seventy-year-old brick machine's rebirth indicated both the resurgence of popularity of soft mud molded brick and the high reputation of the Lancaster Iron Works' Martin Auto-Brick Model 46 brick machine.

A highly reputable French manufacturer (Ceric) installed a soft mud molded brick machine at the Powell and Minnock plant during the late 1980s, but its performance was always problematical. The Powell and Minnock plant was sold in 2002 to an Albany gravel company, and it did not operate in that year. There are no known intentions, of any certainty, to operate it as a brick plant. Barring some presently unforeseen start-up of that plant, this closure also marks the finis of three and one half centuries of brickmaking on the Hudson River.

Notes

1. Related by William Minnock.
2. Three disparate sources have corroborated the account of the critically advantageous southern railways freight rate: William Minnock, Roy Horning (last president of the Lancaster Brick Company), and George Pixly (closely related to the Jova Company of Newburgh). Roy Horning also noted 1966 hearings (conducted in New York City) by the Interstate Commerce Commission with respect to rate reductions for brick shipments from the South. Examination of applicable federal indexes has failed to find ICC hearings specifically relating to such rate hearings. However, such hearings might indeed have been necessary in order to require the connecting railroads (to the north of the southern railway system) to furnish the same low rate into the New York metropolitan area. It would be understandable if the northern railroads had resisted lowering their rates to parity with the southern railroads' rates. Such parity would have been absolutely essential to the success of the southern penetration of the New York City market. Because of the applicability of such hearings to such a narrowly specific subject, they may have been excluded from mention in the federal indexes, or the hearings may only have confirmed a pre-existing requirement for compliance with existing regulations relating to connecting rail rates and their relationship to the rate charged by the originating carrier (in this case, the southern railways).

Principal Sources for Chapter 6

Alecca, Anthony Jr. (See Bibliography).

Haskins and Sells CPA. The Hutton Co. Condensed Reports on Examination for the Year. 1958, 1960.

Hutton Co. *Minutes of the Stockholders and Board of Directors.*

Hutton, GVD. (Statements made to the author).

Minnock, William. (See Bibliography).
Peat, Marwick and Mitchell CPA. The Hutton Co. Financial Statements 1961-1964.
Polacco, John . (See Bibliography).
Powell, Ten Eyck Bronk. (See Bibliography).
Siebert, Robert. *Dennings Point Brick Works 1946-1968.*
Swart, Roger. (See Bibliography).

CHAPTER 7
Epilogue for a Departed Industry

THE HUDSON RIVER BRICK INDUSTRY went down before an array of overwhelming forces, including large demographic changes, competition by new technologies, and by brickmakers in distant locations that had achieved access to the New York market, as well as new environmental standards. During its lifetime, nothing can gainsay that industry's indispensable contribution to the very existence of New York City, where the record of that accomplishment is everywhere to be seen. With the exception of IBM, there is nothing comparable to that industry in the Hudson Valley today in terms of size and consequence of its production.

An Industry's Monuments

One of the great monuments of Hudson River brickmaking, Stuyvesant Town and Peter Cooper Village, was featured in a recent *New York Times* article[1] that stated the intention of the owner-builders, the Metropolitan Life Insurance Company, to continue their enlightened stewardship of this great complex for another fifty years, while making extensive investments in the property at the time of this writing. At the end of that half-century, that complex will be more than one hundred years old. The still bigger Parkchester complex has been made into condominiums after purchase by the Helmsley interests in the late-1970s. That complex, too, will be of great service for a long time into the future.

The American Brick Industry Today

Note should be made of the condition of the brick industry in the United States today. More than one half of the production facilities are foreign-owned. The largest brickmaker in the United States is Boral of Atlanta and is Australian-owned. The third largest, Glen-Gery of Wyomissing, Pennsylvania (that scourge of the Hudson River brickmakers during the housing bureaucracy wars of the 1960s) is Irish-owned. The second largest, Acme of Fort Worth, Texas, has been purposely omitted in the interest of piquing the reader's curiosity and provoking the wildest guess as to the identity of that owner. Given that brickmaking undoubtedly exists in most peoples' minds as some kind of quasi-medieval holdover that is totally irrelevant to the futuristic concerns of our technological world (much less having deserved a penny of any sentient New Economy investor's money), the identity of that owner will come as a very big surprise. It is none other than Berkshire Hathaway and the legendary director of its capital, Warren Buffett. Acme was the victim of internal dissent within its management, a circumstance that undoubtedly created the kind of short-term valuation opportunity that a highly skilled investor with acute hunting instincts would seize. Remembering that Warren Buffett was publicly scorned (and Berkshire Hathaway's share value foolishly downgraded by the market) for his presumed failure to participate in the great technology bubble of the late 1990s, that wise gentleman did have his day after its collapse. Mr. Buffett did some very gracious hand-rubbing, which was addressed to repentant critics: "We have embraced the twenty-first century by entering such cutting-edge industries as brick, carpet, insulation, and paint. Try to control your excitement."[2] Take that, Wall Street.

Echoes from the Nineteenth Century

The recent efforts (summer of 2002) to salvage the turret of the Union ironclad *Monitor* provide a present-day echo from the nineteenth century of the genius of the great steam engine inventor, George Corliss (see Chapter 2 for invaluable contributions to Hudson River brickmaking), whose gear cutting machinery made possible the operation of that turret.

It Was a Very Great Industry

The transfer of appreciable portions of the Hudson River landscape to New York City, in the form of vast amounts of a manufactured construction material—bricks—is witnessed by the inescapable presence of that material in the city. That presence stands as its own monument to the endeavors of the men who did the clay bank excavation by hand, the brick machine crews, the drying yard 'dumpers' and 'hackers,' the 'wheelers,' the kiln setting crews, as well as the entrepreneurs and the men who invented the manufacturing technology for this accomplishment.

The big ponds at Haverstraw and Grassy Point, created by all that intensive brickmaking in the late-nineteenth century, also bear permanent witness to those endeavors and to New York City's dependence on the Hudson River brick industry, as does the little brick church of Our Lady of Mercy at Roseton, which was built for the brickyard workers of that community on the initiative of Juan Jacinto Jova's wife. The new Haverstraw Brick Museum could develop into a fitting memorial to that community's position in the endeavor. Other collections of records and artifacts, as at the Hudson River Maritime Museum (Kingston), should continue to expand and amplify the available record. Hopefully, others will take an active interest in this intriguing subject and amplify the publicly available record with new scholarship and explanation through exhibits and publications.

Silent Witnesses

The silent spirits of the past bear witness also, including the daily group of children who hurried home from school to 'carry the pail' to their hard-working fathers at noontime, the girls with their shy smiles and the boys marching along like little men. Those spirits of the past that will forever inhabit the old centers of production like Haverstraw, Roseton, Fishkill Landing (that name *officially* changed to Beacon), East Kingston, and Glasco include thousands of newly immigrated Irish, Italians, and Hungarians, as well as the workers from Virginia, the Carolinas, and Georgia, and the local worker populations present at the inception of the industry.

Vanishing Traces of Three-and-a-Half Centuries of Brickmaking

Soon enough, virtually all vestiges of the physical presence of all that intensive manufacturing activity will have observably disappeared. The great majority of the substantial changes to the landscape, resulting from the excavation operations, will not be discernible due to overgrowth—a condition that is nearing completion at this writing. Development of brickmaking sites for other purposes will also obliterate all signs of past endeavors. Perhaps one of the two graceful nineteenth-century brick boiler flues at East Kingston (the Shultz yard's flue being a personal favorite), built to power the steam engines that drove the brick machines, will be treasured as the sole remaining industrial artifact of three and one half centuries of the existence of the great Hudson River brick industry.

Notes

1. Oser. "The Upscaling of Stuyvesant Town." *The New York Times*. Jan. 28, 2001.
2. Norris. "Market Place." *The New York Times*. March 12, 2001.

APPENDIX

A Lists of Manufacturers
B Patents Held by Hudson River Inventors
C Brick Prices and Variability
D Principal Publications and Sources of Historical Information on the Brick Industry
E The Maps
F Existing Collections of Artifacts of Brickmaking History
G Productivity: Old Versus New
H Origins of Clay and Its Physical Characteristics
I Mixture Control and Additives
J Scove Kilns—Building and Firing
K Post -World War II Tunnel Kilns
L Kiln Atmosphere
M Kiln Comparisons

A LISTS OF MANUFACTURERS

The following is a survey of lists of Hudson River brickmakers, with comments on completeness and accuracy (see Bibliography and Sources for details on publications).

deNoyelles, Daniel. *Within These Gates*—mostly complete (with exceptions), showing definitive listing of brands; locations and founding dates are erratic. Included in the same publication is a separate list of Hudson River manufacturers that supplied the New York market in 1910.

Hall, C. E. *The Story of Brick*—more limited list than deNoyelles, with same problems of accuracy of locations and founding dates. This list is printed separately.

List of Manufacturers Tributary to New York Markets, 1890—complete listing of makers in the New York City market, with locations correctly named; a field handbook that has brickmaking districts with applicable manufacturers listed on one page and columns for field notes, prices, etc. on the facing page. This list includes manufacturers that were outside of the Hudson River region (Long Island and New Jersey).

NYS Department of Labor. *Second Annual Industrial Directory of New York State*—complete for 1913, with locations accurately named.

U.S. Federal Census. *Products of Industry* —1820, 1850, 1860, 1870, and 1880 are referenced herein; manufacturers are shown under "Bricks and Quarries" heading; great detail provided (in later censuses, particularly 1880) as to value and quantity of annual production, number of employees, wages, type of power, etc.

Ries, Heinrich. *Clay Industries of New York*–New York State Museum Bulletin #12 vol. 3, H. 1895—complete listing that includes identity of market served, machine type, drying system, kiln, and fuel types.

B PATENTS HELD BY HUDSON RIVER INVENTORS

(Listing some of the most important patents)

1828–A. C. Hall (Catskill)—kiln.

1829–James Wood—#97–making brick and tile–3 Mar. 1829, reissued 2 Dec. 1836.

1830–Nathaniel Adams (Newburgh)—brick machine–2 June, 1830, reissued 14 Oct. 1834.

1833–Nathaniel Adams (Newburgh)—tempering wheel–16 Dec. 1833.

1836–Nathaniel Adams (Newburgh)—brick machine–13 April 1836.

1837–Nathaniel Adams (Cornwall)—#376–brick mold and press–8 Sept. 1837.

1845–A. C. Hall (Coxsackie)—British patent # 10,845–brick and tile mold and press.

1847–A. C. Hall (Coxsackie)—#5128–brick press–22 May 1847.

1852–R. A.VerValen (Haverstraw)—# 9,082–brick press–29 June 1852.

1872–R. Lent (Glasco)—#126063–mold sander–23 April 1872.

1893–Charles A. Shultz (East Kingston)—#511098–brick molds (rock face finish).

1902–John Peck (Haverstraw)—process for kiln burning with coal & oil–#708394.

1908–R. C. Penfield (NYC)—brick handling machines (fork lift apparatus for lifting with overhead crane)–#957,368; subsequently in 1910 #962,584; #974,292; #975,713; #976,394; portable brick handling device–#967,663; #976,393

1909–David Strickland (Grassy Point)—#1,012,290 brickmaking machinery–9 Aug. 1909 (assigned to American Clay Machinery Company, Columbus, Ohio); renewed 19 Dec. 1911. David Strickland obtained a total of sixteen U.S. patents, of which seven of the most principal examples are listed herein.

1912–R. C. Penfield (NYC)—brick hacking apparatus–#1,020,572; #1,026,046.

1920–David Strickland (NYC)—#1,344,278 & #1,344,099–automatic brick machines–22 June 1920.

1924–David Strickland (NYC)—#1,495,235–die for brickmaking machines–27 May 1924.

1924–David Strickland—#1,522,600–machine for handling and drying bricks.

1925–David Strickland—machine for handling & drying bricks–pat. # 1,522,600.

1930–David Strickland—#1,772,404–brick kiln–5 Aug. 1930.

RELATED PATENTS:

Friedrich E. Hoffmann (Berlin, Prussia)—#48244 (U.S. patent)–circular kiln (continuous, chamber type)–1865.

C BRICK PRICES AND PRICE VARIABILITY

Sources of Price Information (see Bibliography and Sources for detail on publications):

Engineering News Record–Construction Costs—This highly authoritative publisher shows annual wholesale brick and cement prices f.o.b. NYC market (loaded on carrier at manufacturer's property) for the years 1873–1935 (the only known annual listing of such prices over a period of sixty critical years) and structural steel prices 1894–1935; in same publication is the Construction Cost Index 1903–1935.

Brick–March 1905—spot prices 1870–1904 as submitted by Henry Corse of Glasco, NY.

Clay Record–Jan., May 1908.

Brick and Clay Record–Jan. 1912; Jan. 1928.

Rockland Messenger–1893–Jan.5, July 20, July 27, Sept.14, Nov.30; 1894–May 12, June 28.

NYS Museum Bulletins–*The Mining and Quarry Industry* (series)—#35 (1900), #47(1905), #120(1908), #166(1913), #178(1915), #273(1927), #277(1928).

deNoyelles, Daniel. *Within These Gates*—one early (1860) price and spot prices 1892–1910.

Price Variability: Insufficient identification of prices, with respect to season and location, creates conflicting price data within the same year. Frequently, conflicting quotations may be explained by the time of year at which the price was listed, with spring and late fall prices invariably higher than the mid-season prices at which the largest quantities of bricks were sold. Obtaining an accurate average for annual prices was not feasible, considering that different quantities of brick were sold at different prices. The use of existing information made it virtually impossible to ascertain a weighted average. Even in the most comprehensive list of prices (*Engineering News Record*) 1873–1935, prices are only identified "wholesale f.o.b." in New York City, but not as to time of year. Additionally, spot prices from various sources are frequently not labeled as to their identity as prices at the yard or 'landed' at the NYC dock (add $1.00 to $1.25 per thousand to the yard price for brick landed at the New York City dock). For comparisons, spot prices are therefore generally useful to identify extreme high or low price levels or to indicate trends. In addition, nineteenth-century Downriver bricks could command a price that was ten percent to fifteen percent higher than Upriver bricks. Downright contradictions are also present, as instanced by prices in the NYS Museum's *Mining and Quarry Industries* 1927 issue being twenty percent to thirty percent lower than the *Engineering News Record* prices. In other words, there was no such thing as one price of brick during a calendar year.

D PRINCIPAL PUBLICATIONS AND SOURCES OF INFORMATION ON THE BRICK INDUSTRY

(Complete information on each title is provided in the Bibliography and Sources, listed under author's names).

John McMahon, Dean of the New York State Ceramics College at Alfred University during the early 1950s, noted the meager amount of literature on the Hudson River brick industry (in *A Critical Study of the Manufacturing of Soft Mud Building Brick*). Since that time, the re-publishing and original writing by deNoyelles and O'Connor have appeared, as well as Lawrence and West's writings on ceramic science and Wayne Brownell's writing on New York State clays and shales.

The Story of Brick. C. E. Hall. A state-of-the-art description of the Hudson River Brick industry in 1905, at the zenith of its existence. Besides an adequate general explanation of the manufacturing process, this account describes the operations of fourteen principal yards in considerable detail. An indispensable contemporary account. Available at some Hudson Valley libraries.

Clay Industries of New York (1895), *Clays of New York* (1900), and *History of Clay Working in the United States* (with Henry Leighton). Heinrich Ries. These three volumes are essential to any understanding of turn-of-the-century Hudson River brickmaking, written by a scientist with perception and great interest in the region's industry.

Within These Gates. Daniel deNoyelles. This account, written by a descendant of one of the earliest and greatest of the Haverstraw brickmakers, is essentially limited to Haverstraw and up to World War I; contains much detail and personal explanation. Available at some Hudson Valley libraries and in a current reprint.

A History of Brickmaking in the Hudson Valley. Richard P. O'Connor, PhD (available from UMI Dissertation Services, Ann Arbor, MI)—A doctoral dissertation containing an account of the industry, topically arranged.

Ceramic Science for the Potter. Lawrence and West. An excellent presentation of ceramic science as it pertains to clay, clay-water systems and the process of firing. Written for ceramic arts students, with solid scientific content (derived from the authors' earlier, abstrusely technical work on clay-water systems). This is required reading for anyone interested in this fascinating part of the study. The authors were part of the Ceramics Engineering faculty at the New York State Ceramics College, Alfred University. Richard West was particularly familiar with the Hudson River industry during the 1950s and 1960s. He also served as a consultant for the rebuilding of the Powell and Minnock plant and was an authority on tunnel kilns.

Modern Brickmaking. Alfred B. Searle. A sufficiently detailed exposition of brickmaking facilities and their operation, contemporary in 1911, to give an understanding of a complete range of manufacturing methods. Even though this is an English publication, the manufacturing methods described are essentially applicable to the United States.

A Rudimentary Treatise on the Manufacture of Bricks and Tiles. Edward Dobson similar to Searle (above). Also an English publication, first published in 1850 (with frequent revisions and reprints into the 1930s), it provides considerable information from the early publication date.

Brick and *Brick and Clay Record*. Great amounts of detailed and fascinating information can be obtained from examination of issues of these definitive trade publications, including highly informative equipment advertisements. (*Brick* 1904–1910 is available at the New York City Public Library branch (for Business and Science) at 34th St. and Madison Ave.; *Brick and Clay Record* 1911 and forward is available at both the New York City Public Library and the NYS Library, Albany). Some of the issues are very specifically informative concerning the Hudson River industry, containing essential information that is unique to those publications.

The Clayworker. The same comments regarding the above named journals are applicable in this instance.

Products of Industry. U.S. Federal Census: These highly condensed presentations of essential details of manufacturing enterprises (on the ten-year census cycle, 1820 to 1880) are indispensable sources of information on the region's brickmaking industry; earlier coverage is minimal, with the quantity of detail increasing with successive censuses (1880 far the most comprehensive). Fire destroyed records of the 1890 census and the *Products of Industry* census ceased after that date. This can be found on microfilm copy of original longhand entries, at NYS Library, Albany, or any other library classed as a U.S. Depository.

Reports of the East Kingston Laboratory. Hudson River Brick Manufacturers Association. Contained in a series of notebooks at the Scholes Library (Alfred University) providing a wealth of information of considerable technical content. These reports were written to inform the manufacturers who underwrote the cost of the laboratory, and they directly address their problems and concerns. Not for the general reader.

Clays and Shales of New York State. Wayne Brownell. This includes detailed information and data about the raw materials, written by a top scientist from Alfred University.

E THE MAPS

To date, there have been several maps of brickmaking locations on the Hudson River, each with its own format for indicating the sites of brickmaking. County names are used to identify the principal brickmaking districts on the map entitled *The Hudson River Brick Industry*. Such county identification is used in the lists in the U.S. Census reports and the New York State Museum *Bulletins*. Other lists of manufacturers used various local names to identify the districts—a means of identification employed on certain other maps. The county identification has its own shortcomings. One shortcoming is the case of specific brickmaking localities that are just over the line of a principal county brickmaking district, but so close as to be a part of it. This is true of Coeymans (Albany County), which is just north of the Greene County line. Schodack (Rensselaer County) is in the same category. It is just north of the Rensselaer-Columbia County line.

Both this writing and the maps are concentrated on the largest part of the industry that was supplying the New York City market. Brickmaking locations that were limited to local markets may seem to be disregarded, although the manufacturing in the immediate vicinity of Albany is noted on the map. Another location, such as Arlington (an eastern portion of Poughkeepsie) had a small number of yards for a local market—with the map running out of space to show this kind of exception to the principal New York market.

The location of yards for the Ulster County Brickyards map (unique at the time of this writing) presents problems in terms the difficulty of precisely locating the very large number of yards that existed for a period of more than a century. One of the earliest tasks in this writing was to make an on-site survey of possible locations for brickyards, with the assistance of a U.S. Geological Survey map (such maps label extensive excavation sites as "Quarry"). Considerable brick rubble, bearing one manufacturer's name, spread over a sizable area of the ground and along the river shore circumstantially indicates the past existence of a particular brickyard. However, this assumption can conflict with other information about a specific manufacturer. Some of the lists of manufacturers are also vague as to locations. This latter problem is addressed in the Appendix section "Lists of Manufacturers." The manufacturers that are listed under the category of "Exact Locations Uncertain" are shown for each locality. Specific site locations are so problematical as to require this kind of general listing. However, their existence within those general localities is adequately documented in various noted sources. Specifically located yards, without noted source reference, were identified on the basis of on-site observation of brick rubble and existing structures.

The Haverstraw District map is unique at the time of this writing. The intent is to show the extraordinary concentration of brickmaking along this shore and its permanent effect on the landscape. Evidence on Burleigh's *Bird's Eye View of Haverstraw* and references in Ries's writing in *Clays of New York* help us assume the indicated line of the original face of the bluff running north of Haverstraw (west of the Minisceongo Creek). At any rate, there was extensive excavation from the original bluff at this location. An exact location was admittedly difficult to ascertain.

The *Four Haverstraw Brick Yards* map shows the typical layout of a limited number of

Haverstraw yards, as meticulously delineated on a Sanborn Insurance map of 1896. This map shows the sheer concentration of brickmaking activity that is operating in a very limited area and also contains workers' tenements. Principal features are identified by the Legend in the upper left corner of the map.

The *Bird's Eye View of Haverstraw* of 1884 is fascinating for its detailed and accurate depiction of the manufacturing sites and the landscape. As noted on the illustration, such views present unique records of a landscape that was contemporary with the time of the delineation. The accuracy of this particular representation has been spot checked by comparing specific features on a mid-twentieth century U.S. Geological Survey map and the Sanborn Insurance maps of 1896.

F EXISTING COLLECTIONS OF ARTIFACTS OF BRICKMAKING HISTORY

Haverstraw Brick Museum, Haverstraw, New York: Relates to brickmaking in Haverstraw.

Hudson River Maritime Museum, Kingston (Rondout), New York: Archival materials on local manufacturers and mounted exhibits.

G PRODUCTIVITY: OLD VS. NEW

There has been very little written analysis of the productivity of the Hudson River brick industry, creating a necessity for reasonably careful assumptions and computation when this is done. A recent critic's analysis of productivity has used production data—for the premodernized Jova Company plant—of twenty-eight million bricks per year, manufactured by 325 workers. The analysis then incorrectly hypothesizes that this was done, "during the molding season which lasted slightly more than one third of the year." This assumption was then used as a basis for establishing an annualized production figure for twelve months, for the purpose of equalizing production quantities of the old and new plants. There is no citable evidence for limiting molding to four months. Even relatively early U.S. Census Products of Industry data (1880) show that the production season lasted from six to eight months. Molding was an inseparable part of production. Bricks went directly from molds to drying yards to kilns. The yards could not possibly have operated for six to eight months while molding for only four months. The critic's computation is as follows for the premodernization Jova yard, for annualized production: [28,000,000 bricks (for 4 months) x 3 = 84,000,000/year] divided by 325 men = 258,460 / man / year. The critic's cited evidence (Central Hudson *Bulletin*, Sept. 1926) actually states, "the payroll averaged between 325 and 350 men." The critic selected the lower figure, thereby obtaining a larger production figure per man. The critic rounded off the annualized amount to 260,000/man/year. The four-month operating assumption therefore creates an unexplainably high—and misleading—annualized production rate for comparison with the modernized plant, not to speak of limiting the old yard's payroll to the low end number of workers for the computation.

The critic states annual production for the modernized Jova plant as forty-five million bricks manufactured by 140 workers, at a rate of 321,400 bricks per man per year. The critic identifies the resulting twenty-three percent productivity improvement from modernization (using the critic's selection of a four-month molding season for the old plant) as insufficiently large enough to justify the sizable investment—a foolish waste of money. The implication of unproductive use of capital resources, when applied to Joe Jova (with an RPI engineering degree), couldn't be more inaccurate. The critic states that, "the new (Jova) system was capable of producing bricks regardless of the weather and the season." This assumption, partially correct, is the basis for an unadjusted total production with which to unfavorably compare the premodernization production that had been annualized by using the fictitiously short manufacturing season of four months. The other modernized plants on the river typically operated for no more than nine months of the year (winter shutdown for frozen clay banks and river and for equipment overhaul). The Jova Company apparently experienced a shorter shutdown period (according to the critic's cited reference), thereby requiring at least some upward productivity adjustment, in order to put the modernized plant production on the same annualized (twelve month) basis. Using the critic's only cited data (Central Hudson *Bulletin* of Sept. 1926), the modernized Jova plant had a maximum annual production of fifty-two million bricks—an amount that is not mentioned by the critic, who chose to use the smaller forty-five million amount (stated in the cited reference), which seems to be an average production figure. The cited article states that, "when running at its full capacity it can manufacture approximately 1,000,000 bricks per week." The larger amount isn't nearly large enough to allow for the possibility of a double shift operation—so it must represent the maximum single shift production—and is therefore the correct quantity to use for purposes of annualized productivity measurement: 52,000,000 divided by 140 workers = 371,400 bricks / man / year

Using this productivity amount, and putting both production rates on the critic's annualized basis, provides a revised productivity increase of 44 percent. Nowhere in the cited reference is there any mention of a four-month annual production period. However, the critic's limitation of molding to four months is an erroneous assumption that absolutely requires revision. Since operating seasons ranged between six and eight months, seven months would be a prudent figure to use. An annualized seven-month production season (0.58 year) for the old type of Jova yard would be as follows: (28,000,000 bricks divided by 0.58 = 48,275,860) divided by 325 men = 148,500 / man / year—far smaller than the four-month season selected by the critic (42 percent smaller).

The above adjustment provides a productivity increase of 150 percent for the modernized plant (based on fifty-two million per year)—far different from the figure stated by the critic, with two men (in the modernized plant) doing the work of five men in the premodern plant. Using the rawest comparison (without adjustments for annualization, length of operation during the year, length of workday, etc.) of the basic production and manpower data, the modernized Jova plant (at fifty-two million per year) produced 85 percent more brick with 43 percent of the old yard's manpower (at 325 workers)—a cursory comparison that confirms a large improvement in productivity. Even using the lower production figure selected by the critic (forty-five million) for the modernized Jova plant, the productivity increase is 115 percent—far above the meager 23 percent stated by the critic. Some account must also be made of the critic's selection of the low average number of employees (325) in the old plant for his computation. Using the high average number of 350 workers in the old plant (thereby resulting in lower productivity per worker), the productivity increases for the modernized plant are 170 percent (using new plant production of 52 million bricks) and 130 percent (using new plant production of forty-five million bricks). A realistic figure for productivity increase would fall somewhere between these percentages (115 to 130 percent) for the forty-five million annual new plant production, and 150 to 170 percent for the new plant annual production of fifty-two million. A productivity increase of approximately 150 percent is therefore reasonable (two men in the modernized plant doing the work of five men in the premodern plant). No adjustment has been made for the difference in the length of the workday (twelve hours average in the nineteenth century) because this is unknown for the premodern plant at the time of the recording of the data. The critic's stated productivity increase of 23 percent is actually exceeded by a multiple of five to seven times. Insofar as such productivity increases typified the modernized plants, those modernizations were very good investments by hardheaded men. In the critic's analysis, there were a total of three quite intentional misuses of data—a process that could only be explained by a need to validate a preconceived judgment that had nothing to do with bona fide scholarship.

The foregoing can be verified by using the conventional assumption of an average of one thousand bricks per day per man for the premodernization yard, and by comparing this to the mid-twentieth-century Hutton Company—which was essentially a product of 1920s modernization (see comparison in end notes #6 and #7 of Chapter 5). Wayne Brownell's data from the early 1950s—stating a five hundred million industry brick capacity that was 40 percent of the 1906 maximum production (1.3 billion), manufactured by somewhat less than ten percent of the plants in production in 1906—also restates a significant increase in productivity.

H ORIGINS OF CLAY AND ITS PHYSICAL CHARACTERSITICS

The derivation of clay from feldspar (shown below as a commonly-occurring alkaline type containing potassium), by weathering, in the presence of atmospheric water and carbon dioxide is as follows: $K_2O \,.\, Al_2O_3 \,.\, 6SiO_2$ (feldspar) + $2H_2O$ + CO_2 = $Al_2O_3 \,.\, 2SiO_2 \,.\, 2H_2O$ (Kaolin clay substance) + 4 SiO_2 (silica, as quartz, or sand) + K_2CO_3 (water soluble, being washed out) (Chandler 14). One method of derivation of a Hudson River type of illite clay from feldspar containing potassium is as follows: $3KAlSiO_3$ (feldspar) + 2H (free hydrogen ions) + 12 H_2O = $KAl_3Si_3O_{10}(OH)_2$ (illite clay substance retaining a portion of its potassium) + $6H_4SiO_4$ (silica and water) + 2K (excess potassium) (Ritter 72). Feldspar is a principal component of granite. The silica (SiO_2) and alumina (Al_2O_3) are derived from feldspar and are unaltered during weathering. The silica sand that weathering of the feldspar creates (not incorporated into the clay structure) remains with the clay substance as a separate material. Pure 'clay substance'—kaolinite—has the chemical composition $Al_2O_3 \,.\, 2SiO_2 \,.\, 2H_2O$, with the two water molecules (obtained from the atmospheric weathering of the feldspar) being of hydroxyl type, loosely attached, but within the particle's molecular structure. Thus, there are two very different types of water contained in fully plasticized clay—the external clay-water system that is between the clay particles—and the hydroxyl, chemically bound type within the particles. The external water content between the particles is driven off during the initial 'watersmoking' phase of kiln firing. The hydroxyl water is removed in the 810 degree Fahrenheit to 1,100 degree Fahrenheit temperatures of kiln firing and there is a consequent weight loss of almost fourteen percent (Lawrence and West 115).

Commonly identified as 'clay substance,' clay is comprised of particles that have a diameter of less than two microns (1 micron=1,000th of a millimeter or 0.00004 inch). Particles of this size have enormous surface area relative to mass. Consequently, surface characteristics become highly important—in this case the characteristic of electrostatic charge. It should be noted that there are exceptions to the above stated size limitation, including clay particles as large as one hundred microns. Clay particles can be as small as 0.01 microns. (Lawrence and West 18, 19) Particle shape is very thin, in flat surface dimension relative to thickness (10:1) of irregular hexagonal crystalline outline. These flat crystals are randomly stacked on top of each other. The internal particle structure is also layered, usually having seven planes, with each plane comprised of identical ions of aluminum, oxygen, silica, and hydroxyl layers, forming an interconnected, broad mat. The generic term for such particles (or crystals) is 'layered silicates.' The particles' exterior flat surfaces are surfaced with oxygen atoms, thereby giving those surfaces a negative charge—and also giving the entire particle a predominantly negative charge. Exceptions are kaolin and a limited number of similar clays, having five planes of ions, with hydroxyl ions on one surface. The latter property reduces the plasticity of such clays. (Ritter 66, 80 and Lawrence and West 52). However, the majority of commonly occurring clay types have two negative surfaces, such clays being easily molded (Hudson River clay—illite—is of this type). The space between such particles attracts (adsorbs) positively charged ions (calcium, in Hudson River clays). Water molecules both surround those ions and are electrostatically bound to them (Brady and Weil and Lawrence and West 47–53). A very precise fit exists between the water molecules and clay particles— "the main reason why clay-water systems exhibit their unusual and useful properties." (Lawrence and West 55). A water content that is just sufficient to penetrate the interlayer spaces—both separating the particles and providing a considerable amount of shear-resistant cohesion between them (because of the tight water molecule fit with the clay particle surface)—is called 'water of plasticity' (Lawrence and West 73). In larger increments, the water acts as a lubricant between the particles. Excessive water (in an amount more than barely sufficient to continuously separate the particles) increases the distance between the clay particles, thereby decreasing the shear strength (resistance to rupture by opposite, parallel forces) of the original condition of 'water of plasticity.' Thus, clay becomes an extraordinary compound of tightly associated, dissimilar materials that result in a naturally occurring substance having totally unique characteristics, allowing great freedom in molding followed by transformation into a hard, durable product by the firing process. Water is so integrally bound with the particles that the clay-water system is in a semi-colloidal state—or close to that of a true suspension. The clay particles in a condition of water of plasticity can't just 'settle out.' A true colloidal condition, or suspension, is not quite achieved, insofar as the particles can be separated from the water by the use of a centrifuge.

Hudson River clays are a mixture of related, layered silicate particles, including the predominant type, illite, and chlorites. All layered silicates generally contain the same principal elements, most principally alumina (Al_2O_3), silica (SiO_2) and—in some cases—calcium (anorthite) and potassium (illite).

At the Hutton Company, "quicksand clay" (clay having a significant amount of very fine sand content) occurred in the 'lower bench' (lower level) of the clay bank—a result of the differential, slow settling process of 'lake' types of clays. Transportation of the heavier sand particles were the result of a more rapid glacial melting. Such stratification of the clay banks was common in the Hudson River clay banks. Because of their location to the east of a limestone cliff that terminated a lengthy, east-sloping ledge, Hutton Company clays were among the wettest on the river. Drainage water ran along the top of the ledge and into the clay bank. Wet clays were particularly troublesome in the spring because of the need for added lime to stiffen the mixture (see MIXTURE CONTROL AND ADDITIVES below).

Dry clays were troublesome with respect to the necessary introduction of water for workability. There was a prevalent tendency to introduce excessive water in order to obtain workability as rapidly as possible, followed by consequent problems of excessively soft clay. This could require the addition of calcium (see MIXTURE CONTROL AND ADDITIVES below) in order to stiffen the mix.

I MIXTURE CONTROL AND ADDITIVES

Different plants used different methods to control the clay and sand mixture. In the early 1950s, one leading producer was initially placing sand in the clay cars and adding clay—an imprecise method for control. The Hutton Company had a much more preferable, if more elaborate, system, which included filling separate cars with clay and sand. The sand was then dumped into a separate container adjacent to the machine room (equipped with provision for heating the sand in order to remove excessive amounts of water and to raise the mix temperature in cold weather), from which a feed screw transferred it to hoppers that were positioned above the granulator trench that received the clay from the clay cars. The outlet doors on the sand hoppers had adjustable stops that allowed precise control of the sand volumes, which were then released into the granulator trench. The stop adjustments were made daily, in order to adjust to the clay conditions (Duffy, Polacco).

Excessively wet clays did not hold their shape during the process of removal from the mold. They required the addition of calcium to the mix, in order to stiffen the clay. The water molecules become tightly grouped (attracted by the positive electrical charge) around the calcium ion at the center of each water molecule cluster. The water behaves less liquid (is more tightly structured), and its function as a lubricant between the clay particles is significantly reduced, particularly when the water content is only sufficient to create 'water of plasticity' (see Appendix section H Origins of Clay, above)(Lawrence and West 46–55, 74–75, 78–80). The electrical charges at work are admittedly small (10^6 volts per centimeter on the calcium ion); however, on particles of this size, such charges are strongly attractive for the surrounding water molecules (Lawrence and West 48). The clay thereby loses its excessive softness, becoming much firmer in the molding process. This benefit is offset by the excessive presence of naturally occurring calcium in the Hudson River clay, with its negative effects on firing described herein. Notwithstanding, excessively wet clays (in the spring) were often dosed with calcium. In addition, clays excavated in the spring were also cold, with a consequent lowering of plasticity (workability) (Lawrence & West 77–78). However, William Minnock states that, in practice, this did not noticeably affect the molding process. Heating of the sand removed excess water from the sand and shortened the drying time during the colder spring season, by raising (prewarming) the mix temperature.

Manganese: as an additive to the mixture, manganese yields a brown color, except in the presence of a reducing atmosphere that causes a black color at the brick's surface. As an additive to molding sand, manganese also yields a brown color unless mixed with ferric oxide, thereby yielding a gray-black color.

Coal: generally added by shovelfull measurements.

J SCOVE KILNS—BUILDING AND FIRING

The following Hudson River scove kiln firing information, from various sources is included, with authors' names for identification in the Bibliography.

Ries 1895 (From field observations from early 1890s):
Time for Burning: Wood—five to seven days; oil—four to five days.
Most yards used wood fuel.
Unusable brick: ten to fifteen percent.
Fusibility increases with amount of impurities present, with potash giving highest fusibility, then iron, then lime.

Ries 1900:
Calcareous clays lose twenty percent of their weight in firing.
Final kiln temperature at 1,800 degrees Fahrenheit.
Fine particles expand at 1,000 degrees Fahrenheit, with quartz expanding sixteen percent.
Updraft kilns (scove)—large capacity, cheaper to build, maintain, and operate.
Small amount of fluxible material gives strong hardening with little shrinkage—and still a porous body (Note Ries's identification of this phenomenon, which is specifically identified fifty years later by Richard West).

Brick. January 1904 (Not limited to Hudson River): algamation of inorganic materials into approximately 'perfect' silicate compound, as result of kiln firing.
Change of materials from less stable to more stable condition.
Decarboniztaion of carbonate materials (CO_3 portion of calcium carbonate—$CaCO_3$—changes to CO_2), with the remaining oxygen atom bonding to the calcium (CaO).
Nonvolatile carbon (coal)—burns by slow oxidation—will not burn out until bricks are red hot.
Ferric oxide colors: maximum brilliance at 2,000 degrees Fahrenheit.

McMahon 1948:
Brick becomes lighter, porous (dissociation of limestone—driven off as gas—CO_2).
Critical to maintain high temperature for removal of black cores (in inner portion of brick).
Purer clays melt at 2,068 to 2,291 degrees Fahrenheit, pure kaolin at 3,200 degrees Fahrenheit.
Melt starts at outside of each brick and works inward.
Fired strength—not necessarily increased with higher temperatures.
Initial coal firing for twelve hours of water smoke period.
Black core is a result of coal content and clay properties and is highly injurious to strength; can only be removed by sufficient firing.
Tunnel kilns: more uniform than scove product but not as good as expected—vitrification takes place rapidly (re Hudson River clays).
Coal gases 570 to 1,020 degrees Fahrenheit; clay dehydrates (losing hydroxyl water) 750 to 1,100 degrees Fahrenheit; quartz (silica sand) expands suddenly at 1,060 degrees Fahrenheit; black core removal by complete oxidation at ca. 1,700 degrees Fahrenheit

for ten hours.

Chemical reactions: CaO (Calcium oxide) reacts with alumina and iron ca 1,472 degrees Fahrenheit; CaO reacts with silica at 1,472 to 1,652 degrees Fahrenheit forming new compounds.

Vitrification above 1,800 degrees Fahrenheit: larger amounts of glass production, more shrinkage, lower water absorption, greater strength; Soft mud practice—not to produce a lot of glass *but just enough to glue unfused particles without mass softening* (in the italics phrase, McMahon anticipates Richard West's more specific identification of the function of the flux materials; his implication of moving into clay particle vitrification temperatures is at variance with West, Amberg H.R. Brick Manufacturers Laboratory report of Oct. 1949 and with actual kiln firing practice).

Amberg 1952:

Start of melt stage (vitrification) is abrupt, with large amount of liquid formed over short temperature range.

Hudson River bricks must be fired to temperature just short of vitrification (verifying maximum temperatures that preclude clay particle vitrification).

Carbon and sulfides must be thoroughly oxidized before reaching clay fusion temperature. Any ferrous state iron must become ferric; if not, ferrous iron at interior will become glassy while exterior still below maturing temperature, forming honeycombed interior—lammies—due to ferrous iron oxide's low melt temperature.

A scove kiln is like a chimney of large cross section, with lots of baffling in it.

While a scove kiln may need 90 to110 hours to fire, the rate at which any particular level is fired may be very rapid; too rapid—resulting in incomplete oxidation and formation of damaging amounts of ferrous oxide; too slow—resulting in underfiring and low strength.

Excessive moisture content causes steam that fills kiln's internal air spaces, thereby reducing draft and blocking even distribution of heat.

Rate of fuel feed—too fast will fuse arches; too slow will not provide high enough temperature to heat up the air sucked into arches by draft (for draft maintenance).

Unevenness of hacking—localized quantities are underburned.

Arch damper panels give some control of air movement.

Each arch is essentially a separate kiln.

West 1956:

Dissociation of limestone begins at 1,450 degrees Fahrenheit and is an endothermic reaction, with evolution of large amounts of CO_2, with CaO residue remaining in brick. This is highly reactive at 1,600 degrees Fahrenheit, reacting with alumina and silica to form anorthite and diopside (layered silicates that are closely related to clay substance). These add great strength to brick and also removing lime as soluble substance. Anorthite also reduces firing range of clay (compounding the reduction caused by the calcium).

Limestone (calcite) dissociation removes a large amount of heat at top firing temperatures (endothermic reaction) and heat must be added to avoid underfiring (the function of the brick's internal coal—culm—content); the CO_2 evolved from the limestone dissociation is an oxidant and will burn any carbon content with (exothermic) evolution of heat.

HUDSON VALLEY BRICK MANUFACTURERS ASSOCIATION—LABORATORY REPORTS:

October 1949: Temperature differentials—arches can be at 1,600 degrees Fahrenheit while top levels at 212 degrees Fahrenheit; can have spread of 245 degrees Fahrenheit between top temperatures in different parts, with some parts rapidly getting to finishing temperature and others much slower, with consequent shorter finish temperatures.
Finish (maturing) temperatures—minimum 1,600 degrees Fahrenheit maximum 1,820 degrees Fahrenheit. (Verifying maximum temperatures that preclude true clay particle vitrification.)
$CaCO_3$ (limestone) forms $CaSO_4$ (calcium sulphate) when calcium reacts with H_2SO_4 (sulphuric acid) content in the gases formed by coal combustion, causing kiln scumming (white surface discoloration)—impossible to prevent in upper part of kilns.
'C/B Ratio' (a measure of absorption) and strength are more related to the degree in which carbonaceous cores are burned out at top temperature limits—depending on kiln atmosphere and freedom of circulation of firing gases past bricks.
Black cores leave sunken spots and discoloration at surface.

January 1950: Small black or brown cores preferable to colorless cores, the latter indicating over-oxidized bricks that are weaker (verified in report of March 1951).
November 1950: The short firing range of the clay makes small variations in kiln firing result in large strength variations.

March 1951: Fine capillarity of these (Hudson River) clays makes coal oxidation slower; remedies—use less coal (get less heat); use more sand to open pore structure (sand decreases strength unless fired to higher temperature).

April 1951: With clays having constant moisture content, stiffness increases markedly with temperature decrease; bricks formed at lower temperatures give greater drying shrinkage and tend to vertical laminations.

September 1951: Stiff mud (extruded) bricks, made with a combination of Hudson River clay and shale, did not provide any improvement in the clay's short firing range; they also exhibited high shrinkage at higher kiln temperatures.
Hutton Company clays (three types)—having varying maturing temperature requirements, with one type needing high temperatures for maturing; other types, requiring lower temperatures, to be added into mixture—with such variation requiring thorough mixing of the clay types.
Slow temperature rise to boiling point for release of all hygroscopic 'pore water:' moderately slow rise to 932 degrees Fahrenheit to begin release of all chemically combined hydroxyl water.

Seger 1906 (not limited to Hudson River):
Hydroxyl water is 3.4 percent of dry weight; expelled only by firing.
Calcareous clay: loss of hydroxyl water and loss of carbonic acid (H_2CO_3)—which is 44 percent of calcium carbonate—giving lighter, more porous brick; slight initial shrinkage to beginning of vitrification, then rapid loss of shape, in contrast with high vitrifi-

cation temperature clays that maintain shape, edges, etc.; calcareous clay "Does not stand in fire"—requires skill to avoid fusion; as flux, $CaCO_3$ (limestone) contributes vitreous material that fluxes through mix, lowering fusion temperature and providing vitreous fluxes as filler—but not to exceed 12 to 15 percent of mixture.

GENERAL: Mid-Twentieth Century Scove Kiln Practice:
(Sources: William Minnock and John Polacco)
Kilns were both built and taken down in twelve days (for each operation).

The arch openings in the Hutton Company permanent kiln walls were four feet on center, a common practice with scove kilns, which placed the heat sources at a spacing that allowed reasonably uniform heating of the kiln mass.

Beginning at the floor (hearth) level, ten to twelve courses (the bottom bench) were set. Spaces between the first course of bricks (hard-burned bricks set on edge on the kiln hearth) were filled with pulverized coal (and had the same thermal conservation effect as using 'double coal' brick on walls). This provided the brick mass (for support of the kiln above) on each side of the arch openings. The top courses were corbelled by hand to construct the top of the arch. A forklift truck and manual labor set the bottom bench. (Minnock) The 'boat units,' or blocks of bricks that were then placed in the kiln (1,542 standard bricks, 1272 jumbo bricks), were arranged as layers (benches), to a height of three benches (total kiln height of 56 to 58 courses on edge), and set in the kiln with the overhead crane.

The top covering of the kiln consisted of a single course of double coal bricks on edge, followed by a single course of previously burned double coal bricks (the 'platting') laid flat and tight against each other. Because of the sufficiently irregular shape of the platting bricks, that tight setting did not interfere with draft (air movement through the kiln mass). The access end of the finished kiln (without permanent kiln wall) was mudded over and then braced (shored) with heavy timbers in order to stabilize that vertical wall of brick.

Before the use of overhead cranes and permanent kiln walls, the building of a kiln commenced with the central arch (midway along the kiln's length). This arch was called the 'blade' in some quarters. Further building of the kiln proceeded on each side of the blade with two setting crews proceeding away from the blade toward the ends of the kiln mass. The alternative would have been to start building at one end of the kiln (allowing only one setting crew), thus doubling the amount of time required to finish the kiln. This use of the blade only applied to kilns that were manually built. This practice disappeared with the advent of overhead cranes.

Firing commenced with bituminous coal that was placed near the outer ends of the arches, in order to gradually heat the kiln perimeter. After approximately twelve hours, oil firing commenced. Kiln burner nozzles were initially aimed diagonally down toward the kiln floor, in order to prevent the early penetration of heat to the center of kiln and to localize initial heating along the perimeter of kiln, where the heat loss was highest. (Burner nozzles were mounted on a pipe extending from the main supply line—a 'tee' fitting—with the main supply line running the length of the kiln wall). This initial localization of heat application ('drying out the heads') was essentially similar to practices in use before the advent of oil as a fuel. After the 'heads' had been sufficiently heated, the burner nozzles were rotated up to a horizontal position in order to allow the penetration of heat into the kiln arch tunnel. During the period of increase of the kiln temperature, the outer portions of the kiln had to be kept at a higher temperature than the inner mass. If the inner mass temperature initially got above the temperature of the outer portions, it was impossible to subsequently equalize them. The general effect of this procedure was to raise the temperature of the entire kiln mass as evenly as possible. Subsequently, larger pieces of coal that had been culled out for the purpose were thrown far back into the arch tunnels, beyond the range of the oil burner nozzles (providing heat at the farthest point in the middle of the kiln). This took some skill, in order to propel the coal as far back into the arch tunnel as possible, without hitting either the burner nozzle that was in the same opening or the air duct that ran across the front of the arch opening. A man like Jim Polacco (at Hutton Company.) was good at it and also seemed to enjoy it—like a game. The necessity for this use of coal in the interior of the arch tunnel indicates shortcomings in the oil projection by the pressurized oil fuel system in relation to the length of the arch tunnel (the oil had to be atomized for proper combustion; it could not be a stream).

Firing commenced with No. 2 fuel oil, then switched over to a heavier bunker type of oil ('Bunker C') pressurized at 27 psi (for atomizing) and ducted air at 3 psi, with a return to No. 2 oil for final finishing. At the Hutton Company, there were thirty-one arch openings on each side of the permanent kiln walls, representing the maximum size of the kiln capacity. However, the space between the permanent kiln walls was frequently less than completely filled. Fifteen arches were a workable minimum (sufficient mass was required for economical and even firing).

The hanging gauges (to measure vertical height reduction of the kiln mass), referred to in Chapter 2: Firing A Kiln, were used to assess the rate of temperature change after the second day of firing. The distance between chalk marks (marking the progress of the counterweight) on the board at the side of the kiln were used to compare required standard distances the counterweight was to travel within a specified interval of time. The kiln foreman was responsible for setting the chalk marks and for timing the intervals. The rate of fuel supply was adjusted on the basis of those observations. (Polacco and Minnock) The total vertical dimensional change (the settle) was approximately six inches. To determine when the kiln firing was completed one would observe the platting (surface bricks) on the top of the kiln. This was viewed from a catwalk between the shed columns along the kiln wall. When the platting exhibited a dull red glow, firing was terminated.

Bricks closest to the firing arches could frequently go well above vitrification temperatures into a range that liquefied the clay particles, creating glassy masses of totally fused material that had to be broken up with a sledgehammer. (Polacco)

CONCLUSIONS:

Calcium reduced the firing range to impossibly narrow limits for the scove kiln, precluding entry into a temperature range that would vitrify the clay particles—in excess of 1,820 degrees Fahrenheit with only 35 to 55 degrees Fahrenheit leeway (firing range) to total vitrification. This was the temperature range needed to achieve sufficient clay particle fusion to meet the SW absorption specification. It was out of reach for Hudson River soft mud brick manufacturers using scove kilns. Such a short firing range would have been problematical for any other type of kiln, as well. Cohesion of sand and clay particles therefore had to rely on the calcium oxide (CaO) fluxes that reacted with the alumina and silica to produce anorthite and diopside (such reactions occurred between 1,472 and 1652 degrees Fahrenheit) Such compounds glued the unfused particles together at temperatures below the 1,820 degrees Fahrenheit incipient vitrification temperature for this clay, with considerable resulting porosity. The absolute minimum maturing temperature was 1,600 degrees Fahrenheit with 1,700 degrees Fahrenheit to 1,800 degrees Fahrenheit as an optimum temperature range. Porosity could only be significantly reduced by going above the incipient vitrification temperature, an unacceptable risk with this short firing range type of clay.

Note should be made of a market phenomenon from the late period of the plant's existence—a type of brick that had imperfections that made them highly desirable. These received the name of 'Sanglyn Specials.' Jack Sanglyn, the long-time kiln shed foreman suggested marketing these bricks, which were double coal bricks that were used to cover the exposed end of the kiln mass, just behind the pale bricks on the outer surface that were plastered with clay. (This was limited to the exposed end portion of the kiln mass, with the other three sides surrounded by the permanent kiln walls.) The Sanglyn Specials were distantly related cousins of the lammies, but without the extreme deformation of the latter, having a moderate, horizontal rupture at the mid-height of the brick. The brick otherwise had straight edges and flat surfaces. There also were localized areas of a somewhat bluish hue that gave considerable interest to this type of brick. These brick had a considerable market following (local Kingston example on a house at corner of Farrely and Abbey Streets) Roger Swart (see Bibliography and References) had an old customer from his Hutton Company days humorously asking (over a considerable period of time) for Sanglyn Specials while Roger was employed at Powell and Minnock. They had a following—and they were unique. (See addendum to Appendix J on page 226.)

K POST-WORLD WAR II TUNNEL KILNS

Heat from cooling end of kiln: significant amounts of heat are reclaimed for both drying and preheating of dried bricks. Kiln air cannot be used to dry bricks (such air containing sulfur-bearing combustion products from the fuel oil and the coal); only heated air from reclamation (by ducts) at the cooling end of the kiln (releasing very large amounts of heat) can be used for drying; reclaimed heat, ducted from cooling bricks, is also used for preheating dried bricks that are moving toward the firing zone of the kiln. Drying is initiated with heat ducted from the cooling bricks. Drying is completed with gas-fired heating. Swindell-Dressler discontinued the earlier practice of drawing air across the cooling bricks for the preheating of combustion air (see Chapter 6). The earlier practice transferred too many problematical emissions from the cooling bricks into the firing zone of the kiln, where they had a deleterious effect on the firing (Minnock).

Propulsion of the kiln cars through recent types of tunnel kilns has been furnished by a hydraulic ram that pushes each car of dried bricks toward the firing zone, then retracting to engage the preceding car (a repetition that was repeated in cycles of approximately one and one quarter hours each), i.e. the cars pushed each other through the kiln.

L KILN ATMOSPHERE

Because of elements such as iron oxides in the clay, the presence of sufficient oxygen in the kiln atmosphere must be assured in order to prevent the oxide elements from losing oxygen content (keeping them fully oxidized).

Adequate oxygen supply is essential for combustion of the kiln fuel and of the coal content in the bricks (Lawrence and West 137, 140) if the naturally occurring ferric oxide content of the clay is to be preserved for its red coloring effect and to prevent damaging formations of vitrified cores due to the reduction of ferric oxide (Fe_2O_3) to ferrous oxide (FeO). This causes deformation of the outer brick surfaces. Ferric oxide is refractory (resistant to physical or chemical change at higher temperatures). Ferrous oxide fluxes (melts and flows) at a much lower temperature and also narrows the firing range. The scove brick kiln atmosphere must be kept in an oxidizing (oxygen-rich) condition, as opposed to a reducing (oxygen-deprived) condition that can injuriously remove oxygen from oxygen-bearing compounds in the kiln (such as ferric oxide). This occurs through the subtraction of oxygen from the ferric oxide (Fe_2O_3), forming ferrous oxide (Fe)) that is caused by highly reactive carbon monoxide (CO). This is produced by an insufficient oxygen supply for fuel combustion: $Fe_2O_3 + CO = 2FeO + CO_2$. Ample oxygen is also required for combustion of the pulverized coal (culm) and organic matter in the clay, with an oxygen-poor atmosphere causing reduction of the ferric oxide to the damaging ferrous oxide state due to utilization (actually, theft) of the oxygen content in the ferric oxide by the combustion of the coal content that was added to the clay-sand mixture. Assurance of sufficient oxygen is most critical in the temperature range 662 degrees Fahrenheit to 1,292 degrees Fahrenheit. The oxidation process is most active after reaching 1,112 degrees Fahrenheit. Ferrous oxide is "a powerful flux that reacts with its siliceous surroundings. . . ." (Lawrence and West 140). An assured flow of air must therefore be provided into the scove kiln. This is the function of the round ducts that discharged air into the scove kiln arches. These ducts are connected to a blower.

Control of kiln atmosphere is an essential concern in all ceramic manufacturing processes, using oxidation or reduction as required. Reducing atmospheres (oxygen-deprived) may be utilized, at later stages of firing, for obtaining various color effects on the surface. Tunnel kiln reducing atmospheres could be obtained toward the end of the firing sequence (just before start of cooling) by spraying the bricks with fuel oil (affecting only the bricks at the surface of the kiln car unit. (Minnock).

M KILN COMPARISONS

NOTES TO KILN COMPARISONS:

Note 1: These production figures are not intended to be an absolute comparison of kiln productivity, insofar as this may be impossible without more complete manning data for both Hoffmann and downdraft types of kilns, as well as a lack of other specific information for making precise comparative measurements of productivity. Indications of relative productivity of the kilns will therefore have to rely on citations within the text based on contemporary opinion and circumstantial evidence, and principally in the discussion in Chapter 3 that relates to relative productivity of scove and continuous kilns.

Note 2: Total production in a plant using multiples of this type of continuous kiln would be a multiple of the production of each group of kiln units. In one instance, four kiln units (of sixteen to eighteen chambers) were consecutively arranged in one loop circuit as a single kiln, thereby forming what was said to be the world's largest kiln (Harvey Haigh's kiln as manufactured with the American Clay Machinery Company, for the Bessemer Limestone Company, Youngstown, Ohio, ca. 1910; four hundred feet long,

Kiln Type	Firing Unit	Quantity/ Firing Unit	Firing and Cooling Time	Total Production/ Day (Note 1)	Waste %
Scove (intermittent type)	Arch - total of 8 to 32 per kiln (one million brick average kiln)	30-50,000	8 days- fire and cool entire kiln	Assuming a 1 million brick kiln=125,000 per day	19th c.- 10% to 15% ; 20th c.- 6% to 7%
Hoffmann (continuous type) (Note 2)	Chamber- 16-18 per kiln unit	20-22,000 (1900)	1 day/ chamber	20-22,000 (Note 2)	less than 5%
	(multiples of kiln units	60,000 (1908)	"	60,000	"
	for larger production)	105,000 (1910)	2 days/ chamber	52,5000	"
Downdraft (intermittent type) (Note 3)	Single Kiln Chamber – 16' - 30' diameter	80,000 to 160,000	(1908)16-20 days to heat & cool each kiln	4,000- 10,000	"
Tunnel (continuous type) (Note 4)	Kiln Car	5400 bricks/ car	14-21 cars/ day	Single kiln: 50,000/day- (1927) 75,600 to 113,400/day- (after 1960)	2.5% to 3.0%

with sixty-eight chambers).

Note 3: Multiple kiln installations were typical with downdraft kiln yards because of their limited capacity and relatively slow operation.

Note 4: Adjustment must be made for the advantages of the modern stiff mud tunnel kiln plant that mechanically sets the bricks on the kiln cars at the extrusion machine, thereby eliminating the additional handling required for the racking of the pallets in the dryer, building of units after drying, setting in the kiln and emptying the kiln, as required by the scove, Hoffmann and downdraft kilns, as well as by the early Hudson River tunnel kiln (ca. 1930) that required steam drying of molded soft mud clay bricks. The setting of bricks on a kiln car must be spaced greater than in scove kiln, in order to allow rapid heat penetration through the mass of bricks. In comparison with the 1960s Hudson River soft mud brick plant with a scove kiln, having approximately forty-five production workers, a mechanized stiff mud plant would have approximately ten fewer workers.

SOURCES FOR KILN COMPARISONS: (Reference name cited corresponds with alphabetical listing in Bibliography and Sources)

Scove Kiln: Hall—*The Story of Brick*; Ries—*Clays of New York* 1900; Ries—*NYS Museum Bulletin* #35 vol. #7 1900; Minnock; Polacco; Swart.

Hoffmann (Continuous) Kiln: Ries—*NYS Museum Bulletin* #35 vol. #7 1900; *Clay Record* Mar. 1908; *Brick* January 1910.

Downdraft Kiln: *Clay Record* Mar.1908; *Brick and Clay Record* Sept. 1918.

Tunnel Kiln: *Brick and Clay Record* Oct. 1927; Minnock.

ADDENDUM TO APPENDIX J—SCOVE KILNS

The following is additional kiln construction information added to p.221 for the second printing:

Corbeled bricks over the top of the firing arch opening (for the entire length of the firing arch opening through the kiln mass) were set tight against each other in order to prevent excessive vertical heat flow between the bricks and to encourage lateral heat distribution at the lowest level in the kiln.

Note alternating continuity of courses (all headers or stretchers) across full width of kiln in Figures 23, 24, 49, 77, to maximize the lateral heat distribution through the "finger spaces" between the bricks set in the "three over three" arrangement. Figure 77 shows that this practice was continued with the units that were pre-built in the three-sided cars in the dryers, then set by overhead crane in the kiln (see Figure 72). It was difficult to obtain even diffusion of heat through the scove kiln mass, thereby requiring the use of all possible means for obtaining such diffusion. This included sufficiently close spacing of the firing arches (four feet on center), prevention of localized escape of heat through the top of the firing arch (as explained above), and parallel (hacked) courses for the full height of the "bottom bench" (see lower left corner of Figure 49), compared to the customary "three over three" arrangement in the upper portion of the kiln.

BIBLIOGRAPHY AND SOURCES

Adams, Arthur G. *The Hudson*. Albany, New York: State University of New York Press, 1981.

Alecca, Anthony Jr. Retired president and business representative of Teamsters Union Local #445. Interviewed 2000 (longhand notes in author's file).

American Clay Machinery Company. *Catalogue No. 90*. Bucyrus, Ohio, 1918.

American Institute of Architects, New York Chapter. *AIA Guide to New York City*. Norval White and Elliot Willensky, Editors. New York: Macmillan, 1968.

Amberg, C. R. *The Development of Improved Soft Mud Brick in the Hudson Valley Region*. Thesis for Professional Ceramic Engineer (PhD.) for NYS College of Ceramics. Alfred University, 1952. Manuscript in Archives of Scholes Library. Alfred University.

American Society for Testing Materials (ASTM) *Standard Specification for Building Brick C62-58 and C62-75a*. Philadelphia: American Society For Testing Materials, 1958 and 1975.

___.*Standard Specification—Facing Brick C-216-60 and C216-77I*. Philadelphia, Pennsylvania: ASTM, 1960 and 1977.

Beecher, Raymond. *Under Three Flags*. Hensonville, New York: Black Dome Press, 1991.

Beers, F. W. *County Atlas of Ulster*. N.p.: Walker & Jewett Inc., 1875.

Black, Mary. *Old New York in Early Photographs*. New York City: New-York Historical Society and Dover Publications, 1973.

Brady, Nyle C. and Ray R. Weil *Nature and Properties of Soils*. Upper Saddle River, New Jersey: Prentice Hall, 1999.

Brick. Jan. 1904; Jan., Mar., May, Sept., Oct., Dec. 1905; Jan., Feb., Mar., April, May, June 1908; Jan., Oct. 1910. Chicago: Kenfield Publishing Company.

Brick and Clay Record. May, June 1911; Jan., Feb., June, July, Oct., Nov. 1912; Feb., May, April, 1916; Jan., Mar., June, July, 1918; Jan. 1920; Feb., April, May, July, Oct., Nov. 1927; Jan., July, Sept., Nov. 1928. Chicago: Kenfield-Leach Publishing Company.

Brick Manufacturers Association of New York. *Hudson River Common Brick*. Three volumes cited. Two volumes printed in 1930s and one in early-1950s. Undated.

Brock, Charles. Series of articles on Haverstraw brick industry in *Rockland County Times*. April 2, 9, 16, 1964.

Brown, Arthur. *Brick Drying*. London: The Clayworker Press, 1902.

Brownell, Wayne E. *Clays and Shales of New York State*. NYS Department of Commerce in cooperation with the College of Ceramics at Alfred University, 1951.

Bruce, Wallace. *Panorama of the Hudson*. New York: Bryant Union Publishing Company, 1906.

Burleigh, L. R. delineator. *Haverstraw, N.Y.* (Birds-eye view). Troy, New York: L. R. Burleigh, Publisher. Milwaukee: Beck & Pauli, Lithographers, 1884.

Cantine, Clarissa. Deed for Sale of Property to William Holmes Liber 74, p. 325 1849. Ulster County Clerk's Office.

(The) Carpenters' Company of the City and County of Philadelphia. *Building Early America*. Radnor, Pennsylvania: Chilton Book Company, 1976.

Catskill Daily Mail. Feb.23, 1987 (Greene Co. brick mfg.) Catskill, New York.

Catskill Village and Greene County Directory. Catskill, New York: Fred E. Craigie, printer and publisher 1896, 1901, 1906, 1910.

Central Hudson Gas and Electric Company. *Bulletin* (Reprinted from *Successful Methods*, July 1926) Sept., 1926.

Ceramics Experiment Station. *Monthly Reports.* H. G. Schurecht, laboratory director. Alfred, New York: NYS College of Ceramics 1938, 1939, 1941, 1942. Archives of Scholes Library, Alfred University.

Chandler, Maurice. *Ceramics in the Modern World.* London: Aldus Books, 1967.

Chernow, Ron. *The House of Morgan.* New York: Atlantic Monthly Press, 1990.

Clay Record. Jan., Feb., Mar., April, May, June 1908. Chicago: Clay Record Publishing Company.

Clay-Worker. June 1904. Indianapolis: T.A. Randall and Company.

Clearwater, A. T., editor. "Kingston" by Howard Hendricks. *History of Ulster County.* Kingston, New York: W.J. Van Deusen, 1907.

Cole, Rev. David. *History of Rockland County.* New York: J.B. Beers, 1884.

Commemorative Biographical Record of Ulster County. Chicago: J.H. Beers and Company, 1896.

Condit, Carl W. *American Building.* Chicago: University of Chicago Press, 1975.

Cordts and Hutton. Deed for Purchase of Lands from William and Eliza Shaw. Liber 132, p. 584, 1865. Ulster County Clerk's Office.

Cordts vs. Hutton Company et al. 146 Misc 10 Ulster County Supreme Court, 1932.

___. 241 AD 648 Appellate Division, 1934.

___. 266 NY 399 Court of Appeals and 241 (1934)

___. NY 400 Court of Appeals, 1934.

Cornwall Local. June 25, 1903.

Crossen, Cynthia. "Market Place" ('New York City Rebuilt…'). *The Wall Street Journal.* 4 Sept. 2002.

Davis, F. A. and Company. *Combination Atlas Map of Rockland County.* Philadelphia, Pennsylvania, 1876.

deNoyelles, Daniel. *Within These Gates.* Thiells, New York: Private publishing, 1982.

Dobson, Edward. *A Rudimentary Treatise on the Manufacture of Bricks and Tiles.* First published by J. Weale, London, 1850. Revised by Alfred Searle. 14th edition. London: The Technical Press Ltd., 1936.

Duffy, Francis. Extensive experience in clay bank operations for over forty years, including all equipment. Interviews 2000 and 2001 (longhand notes in author's file).

Dun, R. G. and Company. Credit report manuscripts (for Ulster County manufacturers): 1870–1890. Historic Records Department at Baker Library, Harvard Business School, Cambridge, Massachusetts.

Early Records of the City and County of Albany and Colony of Rensselaerswyck. Translation from Dutch by Jonathan Pearson. The University of the State of New York: 1869.

Engineering News Record. "Construction Costs." New York: McGraw-Hill, 1935.

(The) Federal Writers' Project. *The WPA Guide to New York City.* Copyright 1939. New York: Pantheon Books, 1982.

Fishkill Landing Machine Company *Catalogue.* 7th edition. Fishkill Landing, New York. Undated.

Fishkill Standard. "The Dennings Point Brickyard." 26 March 1881.

Ford, James. *Slums and Housing.* Cambridge: Harvard University Press, 1936.

Friends of Historic Kingston. *A Walking Tour of the Chestnut Street Historic District.* Research and preparation by Lowell Thing.

Gallery Association of New York State. *A Bird's Eye View of New York; Views and Viewmakers of New York State, 1836-1892.* Catalogue for exhibit organized by the New York State Museum. Hamilton, New York: Gallery Association, 1991.

Giddings, Edward D. *Coeymans and the Past.* Saugerties, New York: Hope Farm Press.

Gilbert, Allan S., Richard B. Marrin Jr., Roger A. Wines, and Garman Harbottle. "The New Netherland/New York Brick Archive At Fordham University." *Journal of the Bronx County Historical Society.* Fall 1992.

Gilbert, Alan S., Garman Harbottle, and Daniel deNoyelles. "A Ceramic Chemistry Archive for New Netherland/New York." *Historical Archeology.* 1993.

Green, Frank B. *History of Rockland County.* New York: A. S. Barnes, 1886.

Greene County News. 25 January 1968, Catskill, New York.

Greene County Historical Society. *Journal.* Vol. 12, issue 2. Coxsackie, New York: summer 1988.

Gurcke, Karl. *Bricks and Brickmaking.* Moscow, Idaho: University of Idaho Press, 1987.

Hall, Charles E. *The Story of Brick.* The Building Trades Employers Association Bulletin. July 1905 (reprinted by Daniel deNoyelles in private publishing, 1972).

Harlin, Kevin. *Times Union.* Albany, New York: 23 October 2001.

Hasbrouck, Frank, editor. *History of Dutchess County.* Poughkeepsie, New York: S. A. Matthieu, 1909

Haskins and Sells, Certified Public Accountants. The Hutton Company—Condensed Report on Examination for the Year—1953, 1958, 1960.

History and Genealogy Notes 1700-1938. Vol. I. Greene County, New York.

History of Greene County. J.B. Beers & Company, 1884. Cornwall, New York: Reprint by Hope Farm Press, Cornwall, New York, 1969.

Horne, Field. *Greene County—A History.* Hensonville, New York: Black Dome Press. 1994.

Horning, Roy A. "The Lancaster Brick Co." *Journal of the Lancaster County Historical Society.* Vol. 94 no. 1, 1992.

Hudson Valley Brick Manufacturers Association. *Reports of the East Kingston Laboratory.* C. R. Amberg, director. In cooperation with the NYS College of Ceramics at Alfred University—July 1949 through April 1952 (dates of monthly reports reviewed by author). In archives of Scholes Library, Alfred University.

___. *Special Report for the Hutton Company.* East Kingston Laboratory, 1955. In archives of Hudson River Maritime Museum, Kingston, New York.

Huey, Paul R. "Early Albany Buildings Before 1790." *Albany Architecture.* Preservation League of New York State. Diana S. Waite, editor. Albany, New York: Mt. Ida Press, 1993.

Hughes, Jonathan. *American Economic History.* Scott, Foresman and Company. 1983.

Hutton Company. *Minutes of the Stockholders and Board of Directors.* Vol. I 1897–1926, vol. II 1926–1965. Original documents in author's possession at time of this writing.

___. Deed of Sale of Hutton Company and Lands to American Brick Company. 1899. Copy in author's file.

___. Deed of Purchase of Hutton Company and Lands from Oakleigh Thorne 1900. Original deed in archives of Hudson River Maritime Museum, Kingston, New York.

___. Topographical Map of Clay Bank 1923. Blueprint in archives of Hudson River Maritime Museum, Kingston, New York.

Iorizzo, Luciano. *Italian Immigration and the Impact of the Padrone System.* New York: Arno Press, 1980.

Jacoby, Stephen. Illustration Nos. 60 and 62. *Architectural Sculpture in New York City.* New York: Dover Publications, 1975.

Johnson, William. *The Austrian Mind* (re Hungarian emigration). Berkley: University of California Press, 1972.

Journal of Modern Italian Studies. Vol.4 no.3. Taylor and Francis Ltd., London, and the Istituto Italiano per gli Studi Filosofici, Naples, 1999.

Jova, Henri V., FAIA, FASID. "This I Remember: Random Recollections of the Jova Brickyard and the Devil's Dance Chamber." Essay manuscript in author's file. Atlanta, 2002.

(City of) Kingston, Office of City Engineer. Maps by G. W. Codwise—Ward 5 (Sheet nos. 17–22) 1914.

Kingston City Directory–1890. Poughkeepsie, New York: Kingston Directory Company, Publishers.

Kingston Weekly Leader. (Obituary article for William Hutton.) 6 March 1897.

Kuznets, Simon S. *National Product Since 1869*. New York: National Bureau of Economic Research, 1946.

Lancaster Iron Works. (Brick Machine Advertisement). *The Clay-Worker*. Indianapolis: T. A. Randall and Company, July 1928.

Lawrence, W. G. and R. R. West. *Ceramic Science for the Potter*. Radnor, Pennsylvania: Chilton Book Company, 1982.

Lawrence, W. G., editor with W. E. Brownell, Richard West, and W. O. Williamson. *Clay-Water Systems*. Ceramic Association of New York, Alfred University 1965.

List of Manufacturers Tributary to the New York Market (Field handbook—see description in Appendix). Publisher not named, 1890. In archives of Scholes Library, Alfred University.

Lloyd, Nathaniel. *History of English Brickwork*. London: H. Greville Montgomery, 1925.

Loose, J. W. W. *Notes on Lancaster Iron Works*. Lancaster County Historical Society, Lancaster, Pennsylvania, 1976.

Lowndes, W. S. and D. K. Boyd. *Plain and Fancy Brickwork*. Scranton, Pennsylvania: International Textbook Company, 1950.

Map of Portion of Ulster County Sect. 22 & 23. Watson & Company, 1891.

McCullough, David. *The Path Between the Seas*. New York: Simon and Schuster, 1977.

McKee, Harley. *Introduction to Early American Masonry*. National Trust for Historic Preservation and Columbia University, 1973.

McMahon, J. F., C. K. Kolstad, and H. H. Barnes. *A Critical Study of the Manufacture of Soft Mud Building Brick*. U.S. Department of Commerce, 1948.

___. "A Critical Study of the Manufacture of Soft Mud Brick." Manuscript in archives of Scholes Library, Alfred University, 1948.

Miller, Matt. "Falling Bricks." *The Daily Deal*. New York. 22 Jan. 2002.

Minnock, William. Third generation brickmaker, partner, and president in firm of Powell and Minnock, subsequently vice president in brick manufacturing operations of General Dynamics; retired 1990; More than two dozen interviews and discussions (longhand notes in author's file), 2000, 2001, 2002, 2003.

___. Longhand manuscript recounting market conditions and labor relations after WWII. In author's file, 2000.

Murray, Stewart. *Thomas Cornell and the Cornell Steamboat Company*. Fleischmanns, New York: Purple Mountain Press, 2001.

Museum Bulletins of the State of New York. *The Mining and Quarry Industry* (series). #47 July 1905, #120 July 1908, #166 Aug. 1913, #178 Nov. 1915, #273 Feb. 1927, #277 Sept. 1928, #295 Feb. 1932, #306 April 1936, #319 Feb. 1939, #343 May 1951.

(The) National Cyclopaedia of American Biography. Vol. XLIV. Clifton, New Jersey: 1962.

Newburgh Daily Journal. (Homer Ramsdell obituary notice). 13 Feb. 1894.

Newburgh Gazette. "Court Is to Pass on Sale." (Re auction of Rose Brick Company). 25 June 1919.

Newburgh Hudson Fulton Commission. *Newburgh and Its Environs*. 1909.

Newburgh Sunday Telegraph. (Steam engine article). 23 June 1889.

Newell, Jeanette. *Then and Now in Grassy Point*. Undated. Publisher not named. Local history collection of New City Public Library, New City, New York.

New York Historical Manuscripts, Dutch: Kingston Papers. Translated by Dingman Versteeg. Holland Society of New York and Genealogical Publishing Company. Baltimore, Maryland: 1976.

New York State Department of Labor. *Second Annual Industrial Directory of New York State–1913*. Albany, New York: 1915.

Norris, Floyd. "Market Place." *New York Times*. 12 March 2001.

Nutt, John. *Newburgh*. Newburgh, New York: Ritchie and Hull, 1891.

O'Connor, Richard P., PhD. *A History of Brickmaking in the Hudson Valley*. Dissertation in the Faculty of Arts and Science, University of Pennsylvania, 1987.

Olde Ulster. Vol. 5, Benjamin M. Brink, editor and publisher. Kingston, New York, 1909

Oser, Alan S. "The Upscaling of Stuyvesant Town." *New York Times*, 28 Jan. 2001.

Peekskill Evening Star. (Brick industry article). 18 Jan. 1967.

Peat, Marwick and Mitchell, Certified Public Accountants. The Hutton Company–Financial Statements, 1961 through 1964.

Polacco, John. Extensive experience with manufacturing operations, equipment and kilns. Third generation employee at Hutton Company and with successor owners, 1945–1980. Numerous interviews and discussions (longhand notes in author's file), 2000, 2002, 2003.

Portrait and Biographical Record of Orange County, New York. New York: Chapman Publishing Company 1895.

Powell, Ten Eyck Bronk (d. 1990). Partner and vice president in the firm of Powell and Minnock. Magnetic tape recorded interview; Ellen McHale Ph.D., interviewer, with Phoebe Powell Bender, 1990.

Reynolds, Cuyler, editor. *Genealogical and Family History of Southern New York*. New York: Lewis Historical Publishing Company, 1914.

Ries, Heinrich PhD. *Clays of the United States*. U.S. Government Printing Office, 1903.

___. *Clay Industries of New York*. Bulletin of the NYS Museum. Vol. 3 no. 12, 1895.

___. *Clays of New York*. (Based upon fieldwork performed 1891–1892) Bulletin of the NYS Museum. Vol. 7 no.35, 1900.

___. *Clays of New York*. (Based upon fieldwork performed 1891-1892). University of the State of New York, 1900.

___. *Economic Geology of the United States*. New York: Macmillan Company, 1905.

___. Physical Tests of the Devonian Shales of New York State to Determine Their Value for the Manufacture of Clay Products. Geological Survey of the State of New York (15th report of the State Geologist), 1897.

Ries, Heinrich and Henry Leighton. *History of the Clay-Working Industry in the United States*. New York: John Wiley, 1909.

Riis, Jacob. *How the Other Half Lives*. New York: Dover Publications, 1971.

Ritter, Dale. *Process Geomorphology*. Dubuque, Iowa: Wm. C. Brown, 1978.

Rockland County Times and Rockland Journal News. "Chronicles of Haverstraw 1880–1990."

(Collection of selected articles), April 1902, June 1902, April 1936, October 1942.

Rockland Messenger. (Series of articles on efforts to obtain cooperation of manufacturers regarding production and prices), 14 Dec.1893, 18 Jan., 15 Feb., 5 April, 12 April, 16 Aug., 9 Dec.1894, 6 Feb. 1895.

Rosenwaike, Ira. *Population History of New York City*. New York: Syracuse University Press, 1972.

Rowe, William III. "The Cooper Union." *Iron Architecture in New York City*. The New York State Historic Trust, with The Society For Industrial Technology. New York, 1972.

St. Mary's Benevolent Association. *Bylaws* (published in Italian, with translation of principal portions). Kingston, New York ca.1910.

Sanborn Insurance Maps: Haverstraw, New York, 1887, 1896, 1921, 1931; Catskill 1889; Kingston (North St.) 1899; East Kingston 1887, 1906; Dennings Point 1927. NYC: Sanborn Map and Publishing Company and Sanborn-Perris Company (after 1896).

Scholarly Resources Inc. *Italians to America*. Wilmington, Delaware.

Schoonmaker, Marius. *The History of Kingston*. New York: Burr Printing House, 1888.

Scott, Peter. *Industrial Archeology in Columbia County, NY*. J. M. Kaplan Fund and New York State Office of Parks and Recreation, 1990.

Scott, Robert. *Croton Gazette*. (Articles on Westchester brick industry),10 Aug. 2000, 1 June 2000.

Searle, Alfred B. *Modern Brickmaking*. London: Scott, Greenwood and Son, 1911.

Seger, Hermann August. *Seger's Collected Works*. (Reprinted from articles published in *Brick*). Chicago: Kenfield Publishing Company, 1906.

Shillito, F. W. *Handbook of Corliss Steam Engines*. Bridgeport, Connecticut: American Industrial Publishing Company, 1899.

Siegel, Philip. *A Means of Thoroughly Burning the Double Coal Brick Placed As a Shell Around Scove Kilns*. Bachelor of Science thesis. New York State College of Ceramics 1938. Manuscript in Archives of Scholes Library, Alfred University.

Siebert, Robert. "Dennings Point Brick Works 1946–1968." Essay for Beacon Historical Society (manuscript). 1983.

Smith, Dennis Mack. *A History of Sicily—Vol. II Modern Sicily*. New York: Dorset Press, 1988.

Stokes, Isaac. *An Iconography of Manhattan Island*. New York: Arno Press, 1967.

Structural Clay Products Institute (SCPI). *Technical Notes on Brick and Tile*. "The SCR Brick Wall." Vol . 4 no. 4, New York, 1954.

Sugar, Peter F., with Peter Hanak and Tibor Frank. *A History of Hungary*. Bloomington and Indianapolis: Indiana University Press, 1990.

Swain, Charles. Minority Historian of Greene County. Interview 2001.

Swart, Roger (d. 2001). Administrative assistant at the Hutton Company and for successive owners 1960–1980 and after that year with Powell and Minnock, brickmakers, Coeymans, New York. Interviews and discussions (longhand notes and magnetic tape interview in author's file), 2000 and 2001.

Sylvester, Nathaniel B. *History of Ulster County*. Philadelphia: Everts and Pick, 1880.

Tappen, Christopher. A Map of Certain Town Lands Laid Out for Moses Cantine Being on the Hudson River about Three Miles East from the Town of Kingston. "Performed 1796." In collection of Saugerties Historical Society.

Temple-Balch Institute Center For Immigration Research. *Ship Manifests*. Philadelphia, Pennsylvania: Temple University.

Tuthill, William. "Managers Try to Save State's Last Brick Works." *The Business Review*.

October 22–28, 2001 Albany, New York.

Uhland, W. H. *Corliss Engines and Allied Steam Motors and Atlas* (with translation from German and additional information by Anatole Tolhausen, civil engineer). London: E. & F. N. Spon, 1879.

Uhland, W. H. and Anatole Tolhausen. *Supplementary Volume to Corliss Engines and Allied Steam Motors*. London: E. & F. N. Spon, 1882.

Ulster County. Clerk's Office: Grantor and Grantee records 1859–1896. (Re: land purchases and sales by William Hutton and others).

___. Real Property Tax Service Agency. Tax Maps of the City of Kingston. Numbers 48.084 and 56.028. 1983.

___. Surrogate Court. Matter of Estate of William Hutton. Audit of Estate. 1899.

___. Kingston Commons Maps (folder #335–plate #3)

Ulster County Historical Society. Julia McEntee Dillon Commemorative Art Exhibit. Catalogue raisonné, 1987.

U.S. Bureau of Census. *Historical Statistics* (including Cost of Living Index 1820–1926 by Federal Reserve Bank of NY) U.S. Department of Commerce, 1975.

U.S. Department of Labor. *Consumer Price Index 1913–2000*. Washington: Bureau of Labor Statistics, 2000.

U.S. Federal Census. *Products of Industry*. 1820, 1850, 1860, 1870, 1880. Enumerators' records (longhand) on microfilm. New York States Library.

___. Ulster County Population Statistics. 1870, microfilm. New York State Library.

—-. Manufactures of the United States. Returns 1860, 1870, 1880. U.S. Government Printing Office.

___. *Report of the Manufacturing Industries of the United States*. U.S. Government Printing Office 1895.

U.S. Geological Survey Map: Haverstraw, New York, Quadrangle. Scale 1:24,000, 1967 (revised 1979).

___. Map: Kingston, New York, East Quadrangle. Scale 1:24,000, 1963 (revised 1980).

___. Map: Saugerties, New York, Quadrangle. Scale 1:24,000, 1963 (photoinspected 1978).

U.S. Patent Office. (See Appendix for list of patents obtained by Hudson River inventors for brick manufacturing machinery).

Van Rensselaer Bowier Manuscripts. (Letters of Kiliaen Van Rensselaer, 1630–1643) Translation by Mrs. Alan Strong. Albany, New York: University of State of New York, 1908.

Watson and Company (Map) Ulster County Sections 22 and 23, 1891. In collections of Ulster County Community College and Hudson River Maritime Museum.

Weatherby, Hazel Cantine. Nearly 300 Years of Moise Quantin and Descendants in America. Private publishing, 1982.

Weaver, Charles E. *Clays, Muds and Shales*. Amsterdam: Elsevier Science Publishers B.V. 1989.

West, Richard. "Functions of Coal in the Manufacture of Hudson River Soft Mud Brick." New York State College of Ceramics, Alfred University, 1956. Manuscript in Archives of Scholes Library, Alfred University.

Wilcoxen, Charlotte. *Seventeenth Century Albany*. Albany, New York: Albany Institute of History and Art, 1981.

Wiles, Richard C. The Charles Ranlett Flint Professor of Economics (Emeritus), Bard College, Annandale, New York.

ACKNOWLEDGMENTS

Acknowledgement must also be made of the generous assistance and contributions many people and organizations provided, including those listed below.

Principal acknowledgements:
Phoebe Powell Bender
Francis Duffy
Greene County Historical Society
Elizabeth Gulacsy (Archivist, Scholes Library at Alfred University)
William Minnock
John Polacco
Suzanne Strickland
Roger Swart (d. 2001)
Joan Van Voorhees (Beacon Historical Society)
Richard Wiles

Gratitude for generous contributions to this writing:
Anthony Alecca
Lisa Burke (Archivist, Senate House Historic Site)
Gaetana Ciarlante
Ynez Jova Cline
Carol Connell (Cornwall Public Library)
Stuart Crowell
David Dahlquist (The Clayworks, Des Moines, IA)
Matilda Cordts Davenport
Vincent DeLuca
Janet Dempsey (Cornwall Town Historian)
Dennis Doyle (senior planner, Ulster County Planning Board)
Peter Dwyer
Karlyn Elia (Ulster County Historian)
Martin Fridson (Merrill Lynch)
Guy Gemello
Roy Horning (retired president of Lancaster Brick Co.)
Patrick Hutton
William Hutton (Jr.)
Amanda Cutler Jones
Henri V. Jova
Marie Koestler (Rockland County Genealogical Society)
John Kwaak (Kingston City Engineer)
Allyne Lange (Curator, Hudson River Maritime Museum)
Jack Loose (Lancaster County Historical Society)

Richard Mayone
William H. Meyer (Haverstraw Town Historian)
Roger Mabie
Mark Mahoney (President, Sioux City Brick and Tile Co., Des Moines, IA)
Paula Hutton McKinley
Helen McLallen (Director, Columbia County Historical Society)
Robert Murphy (President, Beacon Historical Society)
George Pixley (d. 2001)
Cynthia Owen Philip
Ten Eyck Powell Jr.
Rockland County Historical Society
Phillip Schaeffer
Robert Siebert
Clifford L. Smith Jr.
St. Mary's Benevolent Association (Jim Polacco)
Herbert Shultz
Barrett Smith, M. S. (Geology), Hydrologist
Thomas Sullivan (Haverstraw Brick Museum)
Charles B. Swain (Green County Minorities Historian)
Craig Williams (Archivist, New York State Museum, Albany)

Libraries:
Particular note must be made regarding the invaluable service to this writing of the following libraries, including the extensive use of the Scholes Library at Alfred University and the absolutely indispensable services of the New York State Library, Albany, as well as the following libraries: Adriance Library (Poughkeepsie), Baker Library (Harvard Business School), Catskill, Cornwall Public Library, Croton Free Library, Coxsackie, Elting Library (New Paltz), Howland Library (Beacon), Kingston Area Library, Newburgh Free Library, The New City Library, King's Daughters Library (Haverstraw), New York City Public Library (Business and Science Branch), Port Ewen, Field Library (Peekskill), Providence Public Library (Providence, RI), Saugerties, Town of Ulster, Toledo-Lucas County Public Library (Ohio).

Archives:
New York State Museum, Albany; Hudson River Maritime Museum, Kingston; Central Hudson Gas and Electric Co., Poughkeepsie.

INDEX